EVOLUTIONARY ALGORITHMS FOR VLSI CAD

EVOLUTIONARY ALGORITHMS FOR VLSI CAD

Rolf DRECHSLER
Albert-Ludwigs-University
Freiburg, Germany

KLUWER ACADEMIC PUBLISHERS
BOSTON / DORDRECHT / LONDON

A C.I.P. Catalogue record for this book is available from the Library of Congress.

ISBN 978-1-4419-5040-6

Published by Kluwer Academic Publishers,
P.O. Box 17, 3300 AA Dordrecht, The Netherlands.

Sold and distributed in North, Central and South America
by Kluwer Academic Publishers,
101 Philip Drive, Norwell, MA 02061, U.S.A.

In all other countries, sold and distributed
by Kluwer Academic Publishers,
P.O. Box 322, 3300 AH Dordrecht, The Netherlands.

Printed on acid-free paper

To Nicole

CONTENTS

PREFACE

In VLSI CAD hard optimization problems have to be solved. Many alternative optimization techniques have been proposed in the past. While some have shown to work well in applications, i.e. the methods have established over the years, while other techniques got lost.

Recently, there is a growing interest in optimization algorithms based on principles observed in nature. These types of algorithms are summarized in the following under the terminology *Evolutionary Algorithms* (EAs). In this book, the basic ideas of EAs are described and the application of EAs in VLSI CAD is studied. Several successful applications from different areas of circuit design, like logic synthesis, mapping and testing, are described. The main focus of the book is not to present the latest developments in these fields. Instead different EA approaches are discussed and alternative integration methods of EAs are described. It turns out that EAs depend on several parameters. The influence of these varying parameters is outlined. By this, the reader who is not familiar with EAs will get a quick start. On the other hand in the book several new aspects of EAs are discussed. This will help people working with VLSI CAD (in industry or academia) to keep informed about recent developments in this area. Thus advanced users (or those who want to become it) can get informations how to integrate EAs.

The book consists of two parts. The first part discusses basic principles of EAs and gives some easy to understand examples. In the second part several EA applications are given. These applications cover a wide range of VLSI CAD and completely different methods how to use EAs are described. Based on these applications the main characteristics of EAs in this area are studied.

The present monograph describes the results of several years research work that has been carried out by the circuit design group and the EA research group at the Johann Wolfgang Goethe-University, Frankfurt, Germany and Albert-Ludwigs-University, Freiburg, Germany. I want to thank all members of these two groups, since they significantly contributed to this book. Parts of this work have been presented as tutorials at *Genetic Programming '97* in Stanford and

International Conference on Genetic Algorithms '97 in Lansing, respectively, and as an invited talk at *International Symposium on IC Technologies, Systems and Applications '97* in Singapore.

My special thanks to Dr. Henrik Esbensen who significantly contributed to this book by writing large parts of Chapter 4 and Section 6.3. Furthermore, I like to thank Dr. Bernd Becker for his support over the past years.

Finally, I want to thank Mike Casey and James Finlay from Kluwer Academic Publisher for their help with preparing the final manuscript.

Rolf Drechsler

drechsle@informatik.uni-freiburg.de

PART I

BASIC PRINCIPLES

1

INTRODUCTION

During the last decades, the complexity of *Integrated Circuits* (ICs) has increased exponentially. In the 1970's a typical microprocessor such as the *Intel 8080* consisted of about 5,000 transistors while in 1995 Intel's state-of-the-art processor *Pentium Pro* contains more than 5 million transistors. To handle the complexity of todays circuits the design engineers are totally dependent on *Computer Aided Design* (CAD), i.e. software tools. The capabilities and limitations of CAD tools have crucial impact on the performance and cost of the produced circuits as well as on the resources required to develop a circuit. Consequently, VLSI CAD is a very important and increasingly growing research area.

In this chapter some basic problems of the *Chip Design Process* (CDP) are pointed out. Since large problem instances have to be handled exact algorithms can not be applied in general and alternative search strategies have to be considered. As one specific method *Evolutionary Algorithms* (EAs) are considered and their properties are outlined. The main areas where EAs will be applied in this book are introduced, i.e. logic synthesis, mapping and testing. Finally, an overview on the contents of the chapters is given.

1.1 MUTUALLY DEPENDENT PROBLEMS

Most problems arising in VLSI CAD are very hard combinatorial optimization problems. Multiple, competing criteria have to be optimized subject to a large number of non-trivial constraints. To handle the complexity, many problems

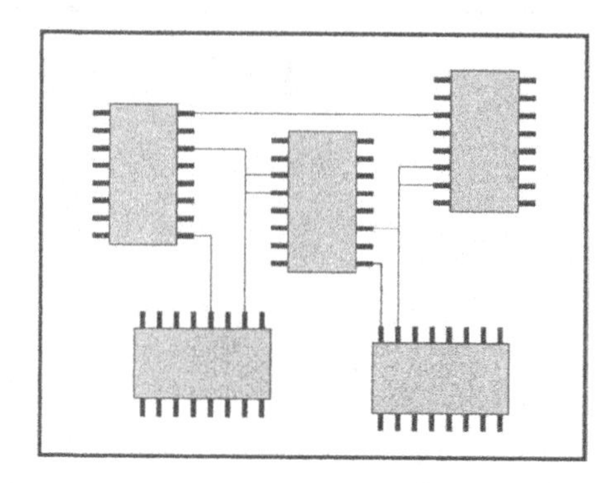

Figure 1.1 Illustration of placement and routing

are artificially divided into a number of subproblems, which are then solved in sequence. Most of the subproblems are still both NP-hard and large, and furthermore, they are mutually dependent. The solution of a subproblem is based on estimations of the results of subsequent steps not yet performed. In other words, the cost functions used are not exact, but are based on estimations.

As an example, consider the problem of placing and interconnecting a very large number of transistors on the two-dimensional surface of a chip. Objectives to be minimized may be the total area, the computation speed of the circuit and its power consumption. Constraints to be satisfied include the physical limitations defined by the technology, specified as minimum distances between transistors, wires, etc. This problem is divided into several subproblems, not all mentioned here. But the transistors are organized in blocks, so that each block implements some sub-function of the required functionality of the circuit. (For illustration see Figure 1.1.) Then, the *placement* subproblem is to place the blocks minimizing the above mentioned criteria. Following placement is the *routing* subproblem, which is the task of implementing wires, which connect transistors belonging to different blocks. Since, in general, wires can not be placed on top of the blocks, sufficient space for the wires has to be allocated between the blocks. I.e., during placement, it is necessary to estimate the space needed between the blocks for the subsequent routing. In other words, the cost function used during placement is based on a routing area estimate. Both placement and routing are NP-hard problems. In fact, just interconnecting a single set of transistors using the minimum length of wire is equivalent to solving the *Steiner Problem in a Graph* (SPG), which is NP-hard. And during routing, typically hundreds, perhaps thousands, of sets of transistors have to be connected. These SPG problems are also mutually dependent. Routes cannot be determined by solving a sequence of SPG problems independently, since the space between any pair of blocks only accommodates a fixed number of wires. This situation illustrates the significant difference between solving an isolated problem such as e.g. the SPG and solving the same problem for a CAD application. In the latter case, a number of additional constraints greatly complicates the problem.

1.2 HEURISTIC MINIMIZATION

Through the years, CAD tools have been developed, based on presumably every existing optimization technique. For example, a variety of deterministic heuristics[1] have been developed for many key subproblems. Formulations based on *Integer Linear Programming* (ILP) are also popular, and so are branch-and-bound algorithms. And if only small instances of a problem occur, simple exhaustive search algorithms may be feasible. By far the most popular stochastic approach in the CAD community is *Simulated Annealing* (SA). This algorithm is known to generate high quality solutions at the cost of excessive runtimes. A vast literature on SA-based CAD algorithms exists, and for several problems, existing SA-based approaches are considered to constitute the state-of-the-art.

1.3 EVOLUTIONARY ALGORITHMS

In recent years an increasing number of EA based[2] CAD algorithms and tools have emerged, offering a realistic alternative to SA. There are (at least) three reasons why the EA should be very well suited for applications in this field:

Complexity: The problems are very hard, as discussed previously, and the relative performance of the EA generally increases with problem complexity.

Parallelism: The EA is inherently parallel. Near-linear speed-ups on MIMD architectures have been widely reported in the literature. This property

[1] In the following the term "heuristic" is used to describe the classical heuristic methods, like greedy, even though the EA is also a heuristic method.

[2] In the following we do not distinguish between GA, GP, ES, etc., and use the unified terminology *Evolutionary Algorithm* (EA).

is especially desirable in the CAD field, where excessive runtimes have always been a well-known problem, and where parallel machines are widely available. The typical IC development site disposes of a large number of interconnected workstations on which parallel EAs can be executed. In contrast, the SA-based algorithm is inherently sequential, and therefore much harder to improve with respect to runtime.

Scalability: When developing a circuit, most steps of the design process are iterated numerous times. This is necessary when it turns out to be impossible to solve a generated subproblem, but iterations are also performed to explore various design alternatives. Therefore, the ideal CAD tool should offer the designer the choice of a reasonably good solution, which can be generated very fast, or a high-quality solution, for which much more time is needed. By nature, the EA has this property. High-quality solutions can be obtained using excessive runtime, but due to the typical very fast convergence of the algorithm during the first few generations, it can also provide a reasonably good solution in a short time.

Although the potential for EAs in VLSI CAD should therefore be large, it should be noted that the EA approach is of interest to the CAD community if and only if it is competitive to the best existing approaches in terms of performance. This has significant consequences, which should be considered when developing EA based CAD tools. Based on this viewpoint, the purpose of this book is:

1. to describe the basic optimization principles of EAs,
2. to describe several EAs in VLSI CAD by reviewing a number of basic algorithms and outlining their recent extensions,
3. to discuss the characteristics of these algorithms, aiming at identifying design principles yielding competitive performance, and finally,
4. to discuss performance evaluation principles. When applying the EA in a practical field such as VLSI CAD, it is crucial that performance is evaluated using the criteria commonly applied in that field, as opposed to traditional EA performance measures.

As one result of reviewing several approaches presented so far it turns out that the major drawback of many approaches is that in general they obtain good results with respect to quality of the solution, but the running times are often much larger than that of classical heuristics.

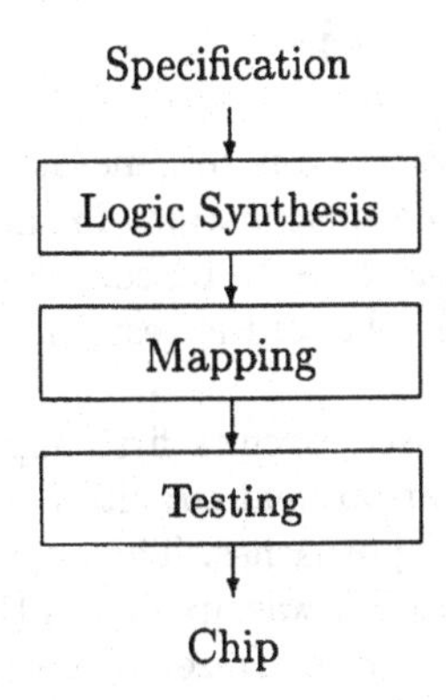

Figure 1.2 Final phase of the CDP

For this finally a new approach to apply EAs to VLSI CAD is discussed: EAs are not directly applied to the problem to be solved. Instead the EA determines a good heuristic with respect to given constraints. The designer himself can for example give upper bounds on the runtime. A model for the description of the learning process is developed. Applications have shown that the model can easily be applied in practice.

1.4 CAD OF IC

The space does not allow a description of the complete CDP. However, in this book only EAs addressing subproblems belonging to the final phase of the CDP are considered, which can be divided into three steps as illustrated in Figure 1.2. Starting from a low-level specification, *logic synthesis* is performed. The optimized circuit is *mapped* to a target technology and finally, the obtained circuit is *tested*. For each of these steps, a number of EA based approaches has been published. Some of these approaches are briefly described in this chapter (without giving any details). The overview in the following is not complete in the sense that all research approaches based on EAs ever published are mentioned. Instead some important ones are picked out, that will be discussed in more detail in Chapter 6.

1.4.1 Logic Synthesis

The main purpose of logic synthesis is to generate a *good* representation of a given function as defined by the specification. In this context *good* means good with respect to constraints that have to be considered, e.g. constraints on the types of basic gates or on the delay of the resulting circuit.

For two-level AND/EXOR based circuits first approaches were presented in [6, 40]. In this application near-optimal circuits were generated and proven to be superior to other heuristic approaches. The approach has been extended in [45]. For multi-level circuits an EA was applied in [101] to learn parameters of heuristic minimizers, like MIS or BooleDozer. An other approach for multi-level synthesis based on mapping *Decision Diagrams* (DDs) to *Field Programmable Gate Arrays* (FPGAs) has been presented in [44]. The approach has been extended in [42] to also consider the testability of the resulting circuit.

1.4.2 Mapping

The generated circuit description is then mapped to a target technology. For FPGAs based on *Configuration Logic Blocks* (CLBs) this has been considered in [92]. In some layout styles, e.g. macro-cell layouts, there is a need for *placement* and *routing*. Much work has been done in this area using EAs [107, 60, 59, 52, 58, 78, 104, 76]. In the following we distinguish between placement, global routing, floorplanning and detailed routing.

1.4.3 Testing

Even if a circuit is correctly designed, a fraction of the produced circuits will be faulty because of physical defects caused by imperfections during the manufacturing process. Test pattern generation, i.e. generation of a test for a given fault, for sequential circuits has been considered in [134, 135, 30, 27, 84, 77]. To reduce the testing time the number of test patterns needed is minimized by test set compaction. A first approach using EAs is given in [133].

Recently, EA-approaches have also been applied in other testing areas, like partial scan flip flop selection [29] and build-in-self-test [121].

All in all, the overview given above briefly describes the areas in the CDP, where EAs have been used. It turns out that EAs have been applied to completely different problems. They are a very general optimization method, but by a more detailed study in the following it will be shown that a "careful" tuning towards the problem considered is needed. This especially means that the EA in its general form is of little use in practice, while the incorporation of problem specific knowledge, e.g. in form of heuristics, makes EAs a powerful tool.

1.5 OVERVIEW

The book consists of two main parts: Part I (Chapter 1-4) introduces the basic concepts, while Part II (Chapter 5-7) is practical oriented. Each chapter starts with a brief introduction to motivate the topics that will be discussed. At the end of each chapter the main conclusions are summarized. By this structure each chapter is (more or less) self-containing. The book is structured as follows:

In Chapter 2 the biological background of EAs is briefly described. Then the classical EAs are introduced. Alternative representation forms are discussed and examples of genetic operators are given. Several special cases of the basic EA concept are mentioned, like evolution strategies and genetic programming.

The "restricting" factors of EA applications, e.g. large problem instances, in the CAD area are considered in Chapter 3.

Performance evaluation for EAs is discussed in Chapter 4. A model for multi-objective optimization is presented that allows to compare elements or sets of elements. A first application demonstrates the quality of the model.

In Chapter 5 the implementation of a software package for EAs with special emphasis on VLSI CAD is described. It turns out that the integration of problem specific heuristics is a key issue of successful applications. For this a hierarchical class concept is developed that allows easy software reuse and combination of different CAD tools.

Several applications from the areas of logic synthesis, mapping and testing are described in detail in Chapter 6. Dependent on the application different representations should be preferred. Methods and variations of the basic EA concept are presented. This gives a better insight in the optimization principles.

Chapter 7 proposes an alternative optimization strategy that is also based on EAs. But the EA is not directly applied to optimize a single problem instance. Instead the EA is used to develop a heuristic for a problem.

Finally, the main results of the book are summarized in Chapter 8.

2

EVOLUTIONARY ALGORITHMS

2.1 INTRODUCTION

In this chapter the basic underlying ideas of EAs are described. First, a brief description of the biological background is given. Then the basics on an implementation on a computer system are discussed. The main components of an EA are presented, i.e. fitness function, selection principle and evolutionary operators. Finally, some special cases of the concept defined here are outlined, like evolution strategies and genetic programming.

2.2 BIOLOGICAL BACKGROUND

We start with introducing some notations and pointing out the analogies between the "real world" and the "simulation on computer systems" that will be used in the following as a minimization principle:

- EAs are derived from observations in nature where living beings are improved by evolutionary mechanisms.
- Each solution is denoted by an *individual* which is represented as a string over a fixed alphabet (*genotype*).
- A set of individuals is called a *population*.
- To each individual in a population a *fitness* is assigned based on an *objective function*. The objective function measures the quality of the solution corresponding to an individual.

- Analogously to nature the individuals in a population can reproduce themselves. This is simulated by recombining genotypes of selected parent elements.
- An element is selected proportional to its fitness, thus individuals (solutions) of higher quality are preferred during the selection process. The selection process in the real world is based on the principle of "survival of the fittest", i.e. the living beings that are well assimilated to their environment have a higher chance of surviving.
- In "classical" EAs the recombination is done by a method called *crossover*. This is based on observations in nature where two individuals are splitted in two parts at cut points. Then two offsprings are created by joining together the parts.

Notice that the main idea when applying EAs for VLSI CAD is **not** to copy nature as good as possible. Instead the underlying principles should be used to get high-quality results.

2.3 GENETIC ALGORITHM AND EVOLUTIONARY ALGORITHMS

In this section the basic ideas of *Genetic Algorithms* (GAs) [83] are described.

Remark 2.1 Sometimes the term GA instead of EA is used here to underline that a restricted optimization method is considered. All GA operators used here are also supported by EAs.

Then some generalizations used in EAs are given. The main components of GAs and EAs are discussed, i.e. representation, objective function, selection method, initialization of population, genetic operators. Based on these components the overall structure and information flow of GAs and EAs is described.

2.3.1 Representation

The standard GAs make only use of binary representations [79]. Thus, each element of the population corresponds to an n-dimensional binary vector. A population is a set of vectors. The size of the population does not have to be determined in advance. It is also possible to dynamically enlarge or reduce the number of elements. (For simplicity we assume a population of constant size in the following if not mentioned otherwise.)

Since in various applications it turned out that other encodings should be preferred, since they are e.g. more problem specific or allow to easily incorporate problem specific heuristics, EAs have no restrictions on the representation form.

Dependent on the encoding method it is also possible that some elements represent infeasible solutions. Here two main alternatives exist:

1. Avoiding invalid solutions by changing the encoding accordingly.
2. Introducing a repair operator that "corrects" the elements in a validation run.

Notice that the validation run is not necessarily needed in each step. EAs can also work on invalid solutions (see e.g. Section 6.3.5). In these cases the fitness function (see below) should "punish" these elements.

2.3.2 Objective Function

The *objective function* measures the *fitness* of each element. (For this it is also called the fitness function in the following.) The choice of this function is very important for the overall quality of the EA. For "simple" optimization problems, it is often straightforward to determine the quality of the solution:

Example 2.1 The number of colors during graph coloring directly corresponds to the result of each element. But notice that such a straightforward measure does not incorporate any detailed information, like e.g. lower bounds on the number of colors.

2.3.3 Selection

The selection method determines the parent elements for the genetic operators. Many different principles have been suggested in the past few years. Here only three are mention that are frequently used in VLSI CAD:

Random selection: The elements are selected at random. (The fitness is not considered.)

Tournament selection: Elements are randomly selected and the ones with the better fitness are used.

Roulette wheel selection: Elements are selected proportional to their fitness.

Since the genetic operators create new elements also a replacement strategy must be considered to limit the size of the population.

Often not the pure methods are used, instead the basic principles are combined with other approaches, e.g. often EAs make use of *elitarism* [34]:

Elitarism: Some of the best elements of the old population are included in the new one anyway. This strategy guarantees that the best element never gets lost. Furthermore, often a faster convergence is obtained.

2.3.4 Initialization

The starting point of a GA is of large importance. Classically, at the beginning of each GA run an initial population is randomly generated and the fitness is assigned to each element. Often it is helpful to combine GAs with problem specific heuristics [34]. The resulting GAs are called *Hybrid GAs* (HGAs). (Analogously *Hybrid EAs* (HEAs) are defined.)

Here the encoding method plays an important role: If also invalid solutions might occur by the representation style it might be that randomly generated elements can not determine a valid starting point. In these cases it is often helpful to use problem specific heuristics at least during the initialization. On the other hand these heuristics should not be used too frequently, since they may force a too fast convergence. By this the power of the GA would be reduced.

2.3.5 Genetic Operators

In the following some basic genetic operators are described. All operators are directly applied to binary strings of finite length that represent elements in the population. (Notice that it is straightforward to apply these operators to multi-valued strings.) The parent(s) for each operation is (are) determined by the mechanisms described in Section 2.3.3.

Reproduction: Copying strings according to their fitness.

Crossover: Construction of two new elements x_1 and x_2 from two parents y_1 and y_2, where the first (second) part of x_1 (x_2) up to a randomly chosen cut position i is taken from y_1 and the second (first) part is taken from y_2.

2-time Crossover: Construction of two new elements x_1 and x_2 from two parents y_1 and y_2, where the first (second) part of x_1 (x_2) up to a cut position i is taken from y_1 (y_2), the second part up to a second cut position $j > i$ is taken from y_2 (y_1) and the last part is again taken from y_1 (y_2).

Universal Crossover: Construction one new element x from two parents y_1 and y_2. At each position the value for x is randomly chosen from y_1 and y_2, respectively.

Mutation: Construction of a new element from a parent by copying the whole element and randomly changing its value at mutation position i.

2-time Mutation: Construction of a new element from a parent by copying the whole element and randomly changing its value at mutation positions i and j $(i \neq j)$.

Mutation with neighbour: Construction of a new element from a parent by copying the whole element and randomly changing its value at mutation positions i and $i + 1$.

It is straightforward to extend *2-time Mutation* to *k-time Mutation*.

For illustration a brief example is given for each of the operators presented above (except *reproduction*) on two-valued strings of length 3:

Example 2.2 1. Crossover: $y_1 = 001$, $y_2 = 110$ and $i = 1 \rightarrow x_1 = 010$ and $x_2 = 101$

2. 2-time Crossover: $y_1 = 001$, $y_2 = 110$, $i = 1$ and $j = 2 \rightarrow x_1 = 011$ and $x_2 = 100$

3. Universal Crossover: $y_1 = 001$ and $y_2 = 110 \rightarrow x = 000$

4. Mutation: $y = 001$ and $i = 1 \rightarrow x = 101$
5. 2-time Mutation: $y = 001$, $i = 1$ and $j = 3 \rightarrow x = 100$
6. Mutation with neighbour: $y = 001$ and $i = 1 \rightarrow x = 111$

2.3.6 Algorithm

Using the operators introduced above a "classical" GA works as follows:

1. Initially a random population of binary finite strings is generated.
2. The genetic operators reproduction, crossover, and mutation are applied to some elements. The elements are chosen according to the selection method.
3. The algorithm stops if a termination criteria is fulfilled, e.g. if no improvement is obtained for a fixed number of iterations.

The GA depends on several parameter settings, e.g.

- population size,
- reproduction probability,
- crossover probability,
- mutation probability.

A sketch of a "classical" GA is given in Figure 2.1.

2.4 EXTENSIONS OF THE CONCEPT

Beside GAs as described above several alternative concepts have been proposed, e.g. *Evolution Strategy* (ES) [128, 4] and *Genetic Programming* (GP) [93]. These approaches differ from GAs mainly in the way of representing elements, e.g. GPs represent elements in a tree-like structure, and ESs allow representation of

```
genetic_algorithm (problem instance):
    generate_random_population ;
    calculate_fitness ;
    do
        apply_operators_with_corresponding_probabilities ;
        calculate_fitness ;
        update_population ;
    while (not terminal case) ;
    return best_element ;
```

Figure 2.1 Sketch of GA

elements by not only binary but real numbers. Furthermore, the selection principles slightly differ.

In this book we focus less on the differences between the alternative approaches. Instead we are interested in constructing powerful VLSI CAD algorithms. For this the term *Evolutionary Algorithm* is used in the following to summarize the different approaches.

2.5 SUMMARY

In this chapter a brief description of EAs has been given. The main underlying ideas have been outlined and the notation and terminology that will be used in the following chapters has been introduced.

3

CHARACTERISTICS OF PROBLEM INSTANCES

3.1 INTRODUCTION

The requests on algorithms applicable in VLSI CAD differ from other areas. This also influences the design of EA based tools. For this, in this chapter some important aspects are highlighted that have to be considered.

3.2 SIZES OF PROBLEM INSTANCES

As described in the previous section EAs are based on measuring the quality of an element in the population. The larger the population is the more elements have to be evaluated. In areas like logic synthesis this can be very time consuming, especially if several optimization criteria, like area, delay, power and testability, are considered in parallel.

Nowadays chips may contain thousands of gates and transistors, respectively. For this, often only "simple" estimation methods can be used on the higher design levels within reasonable time bounds. If also other parameters, like e.g. technological constraints, have to be considered the situation becomes even worse.

Since EAs are mainly simulation based, in many applications the evaluation of the objective function takes more than 99% of the overall runtime. For this, it is also reasonable to modify the fitness function during a program run. First, only rough estimations are used, while during the final optimization more time consuming methods can be applied.

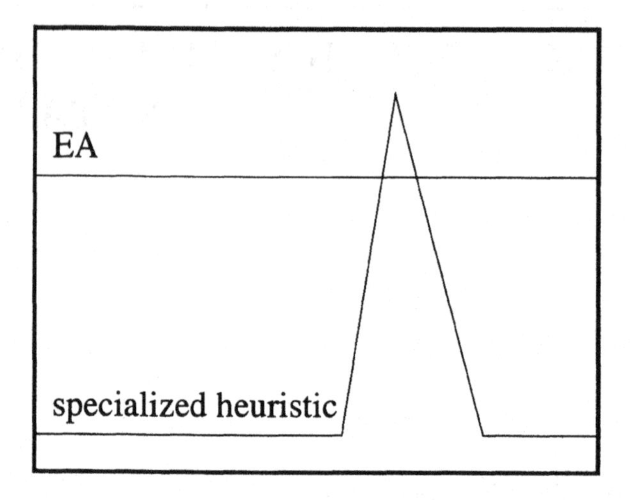

Figure 3.1 Performance of "typical" EA

3.3 QUALITY-SPEED TRADE-OFF

In general (assuming NP not equal to P [72, 116]) most exact algorithms for "hard" problems can only be applied to small problem instances. For larger instances heuristic solutions are often needed.

This can be "loosely" summarized as:

- Fast algorithms obtain "bad" results.
- Time consuming algorithms obtain "good" results.

The point that is addressed in the following is how the "typical" EA can be rated in this simple scheme. In Chapter 1 in [79] the "typical EA performance" is described as a robust algorithm, that will be outperformed by some highly specialized method. This is shown in Figure 3.1 where the different problems are given along the x-axis. Several studies showed that this is mainly true, if the EA does not make use of problem specific knowledge. In this case the EA principles are of only minor usefulness to the CAD community. But in the following (see Chapter 6) several examples will be given that demonstrate that a clever combination of existing heuristics with EA principles can often give much better results with respect to quality of the result. As an other aspect also the runtimes of the algorithms have to be considered. It is otherwise not surprising to obtain very good results using "arbitrary" runtime.

The main advantage of an EA in this context is that by varying the parameter settings and integration of problem specific algorithms, the wide gap between exact algorithms and simple greedy heuristics can be closed smoothly. If for example a small population is used and an "early" termination criteria the EA runs faster, but the results are probably less good.

3.4 SUMMARY

In this chapter we mentioned some aspects that should be considered if EAs are applied to VLSI CAD. By this, some important points have been highlighted that will be addressed more detailed during the descriptions of the applications later.

4

PERFORMANCE EVALUATION

4.1 INTRODUCTION

Before focusing on applications of EAs this chapter describes some important aspects that have to be addressed when working with evolutionary techniques. The problems described in the following are not all that have to be considered, but the discussion points out essential parts.

First, the problem of measuring the quality of a solution for an optimization problem is highlighted by an example. Then an efficient method for design space exploration from [55, 56] is presented. Some advantages of generating solution sets rather than single solutions are presented. Finally, the trade-off between quality and speed during the optimization process is discussed.

4.2 MEASURING PERFORMANCE

Before studying a method for design space exploration in more detail the difficulties of measuring the quality of a solution candidate will be shown by a simple example (see [112]):

Example 4.1 Let A and B be two points that have to be connected by a line within a rectangular region. The connection line is not allowed to go through any of the dark boxes. Furthermore, the line should have a minimal length.

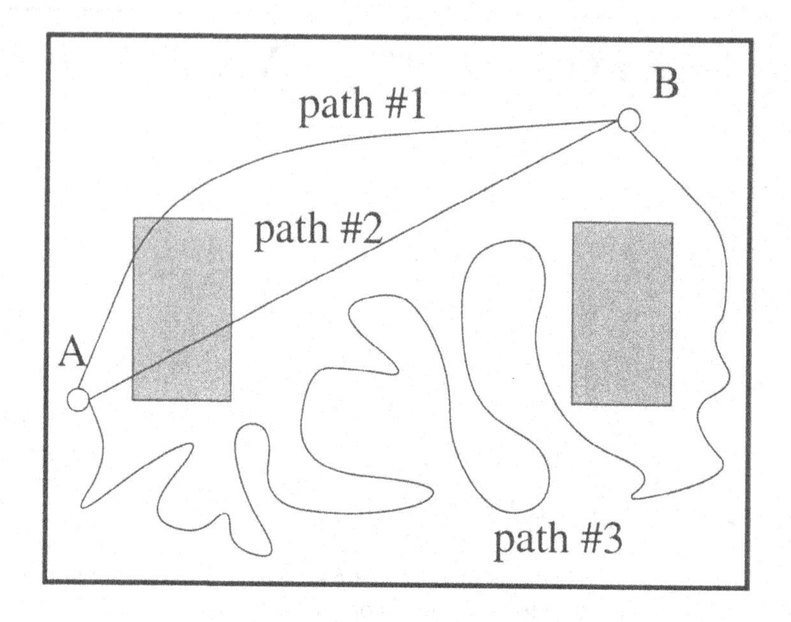

Figure 4.1 Finding the minimal path

In Figure 4.1 a problem instance is given and three lines, denoted by path #1, path #2, and path #3, are given:

Path #2 is a trivial "solution". It optimizes the length in the way that it is the minimal solution, but it is not valid, since one box is touched.

Path #1 describes a "better solution", since it at least tries to avoid the obstacle, but since it still touches the box it is also not valid.

Path #3 is the only line that fulfills both criteria, i.e. A and B are connected and no box is hit. But the solution is far from being optimal, since the line is very long.

Even this simple example shows that it largely depends on the problem to be solved whether invalid solutions should be considered at all. Therefore, techniques are needed that allow comparison of two solution candidates.

4.3 DESIGN SPACE EXPLORATION

The objective of virtually all multi-objective optimization problems arising in VLSI CAD (and many other combinatorial optimization problems as well) is to find a solution which is satisfactory with respect to a number of competing criteria. Most often specific constraints have to be met for some criteria, while for others, a good trade-off is wanted. The traditional problem formulation used to obtain this objective is to minimize a weighted sum of some criteria subject to constraints on others. I.e. if k criteria are considered, the objective is to minimize the scalar-valued cost function

$$c = \sum_{i=1}^{j} w_i c_i \quad \text{such that} \quad \forall\ i = j+1, \ldots, k : \ c_i \leq C_i \tag{4.1}$$

for some j, $1 \leq j \leq k$. Here c_i is the cost of the solution with respect to the ith criterion and the w_is and C_is are user-defined weights and bounds, respectively.

However, for VLSI CAD problems, the expected values of the cost criteria are based on relatively rough estimations only, especially in the early phases of the design process. Furthermore, the available information on obtainable trade-offs, for example the relationship between area and delay during floorplanning, may be very limited or non-existent. Since the notion of a "good" solution inherently depends on which trade-offs are actually obtainable, the overall design objective is rarely clearly definable. Consequently, it may be very difficult to specify a set of weights and bound values that makes a tool based on the formulation of Expression (4.1) find a satisfactory solution.

Even when assuming a clear notion of the overall design objective, the use of Expression (4.1) causes serious difficulties:

- If the bounds are too loose, perhaps a better solution could have been found, while if they are too tight, a solution may not be found at all.
- Finding suitable relative weights may also be very difficult in practice. If some criterion is not optimized sufficiently well, it is most often impossible to know a priori how much the corresponding weight should be increased.
- Different characteristics of the individual cost functions c_i may make constant weights insufficient to keep the cost factors $w_i c_i$ properly balanced throughout the optimization process.
- The minimum of a weighted sum can never correspond to a non-convex point of the cost trade-off surface, regardless of the weights [63]. In other

words, if the designers notion of the "best" solution corresponds to a non-dominated, but non-convex point, it can never be found using Expression (4.1).

Some recent approaches for various VLSI CAD problems address these problems [25, 33, 108, 58]. They explicitly explore the design space and generate a set of alternative solutions representing distinct, good trade-offs, from which the user can then make a final choice. However, a fundamental but still unsolved problem is how to properly compare the performance of design space exploration approaches [65]. Such comparisons requires a method for comparing *sets* of solutions, all of which represent "good" trade-offs of the cost dimensions.

In this section a preference relation on multi-dimensional cost spaces is introduced, allowing individual solutions to be compared without resorting to the use of a single-valued cost functions such as Expression (4.1). Furthermore, a solution set quality measure is proposed which allows set-generating algorithms to be evaluated. The measure is independent of the search algorithm used and therefore applies to the evaluation of any set-generating approach, for example a random walk, an exhaustive search, or approaches based on simulated annealing or EAs. It can also be used to measure the progress of algorithms based on iterative improvement of (one or more) solutions, such as the EA.

In the following a relation on the cost space allowing individual solutions to be compared is first presented. Then the solution set quality measure is introduced and some properties of the measure are proven. The practical application of the measure is discussed and an empirical validation using constructed sets follows.

4.3.1 Comparing Single Solutions

Let Π be the finite search space considered and let n be the number of distinct optimization criteria considered. Assume without loss of generality that all criteria are to be minimized. The cost measure is the vector-valued function $c : \Pi \mapsto \mathbf{R}_+^n$, $c(x) = (c(x)_1, \ldots, c(x)_n)$, where $\mathbf{R}_+ = [0, \infty[$.

As mentioned above, the traditional way of combining the $c(x)_i$ values into a single-valued cost measure using weights and/or bounds causes some practical problems to be avoided. Therefore, a different approach is taken. An ordering

$\prec$ on Π is introduced, which allow individual solutions to be compared without aggregating the cost values.

Let

$$\mathbf{R}_{+\infty} = [0, \infty]$$

and

$$G = \{(g, f) \in \mathbf{R}^n_{+\infty} \times \mathbf{R}^n_{+\infty} | \, \forall i : g_i \leq f_i\}.$$

Instead of weights and bounds, the user specifies preferences by defining a *goal and feasibility vector pair* $(g, f) \in G$. For the ith criterion, g_i is the maximum value wanted, if obtainable, while f_i specifies a limit beyond which solutions are of no interest.

Example 4.2 If the ith criterion minimized by an IC placement algorithm is layout area, $g_i = 20$ and $f_i = 100$ states that an area of 20 or less is wanted, if it can be obtained, while an area larger than 100 is unacceptable. Areas between 20 and 100 are acceptable, although not as good as hoped for.

Since the values of (g, f) need *not* be obtainable, in contrast to traditional bounds, they are significantly easier to specify.

For $(g, f) \in G$, let

$$S_g = \{x \in \Pi \mid \forall i : c(x)_i \leq g_i\}$$

and

$$A_f = \{x \in \Pi \mid \forall i : c(x)_i \leq f_i\}$$

be the set of *satisfactory* and *acceptable* solutions, respectively, as illustrated in Figure 4.2.

Remark 4.1 $S_g \subseteq A_f \subseteq \Pi$ for all $(g, f) \in G$.

Finally, the notion of *dominance* is needed.

Definition 4.1 Let $x, y \in \Pi$. The relation x *dominates* y, written $x <_d y$, is defined by

$$x <_d y \quad \Leftrightarrow \quad (\forall i : c(x)_i \leq c(y)_i) \land (\exists i : c(x)_i < c(y)_i)$$

An ordering on Π can now be defined:

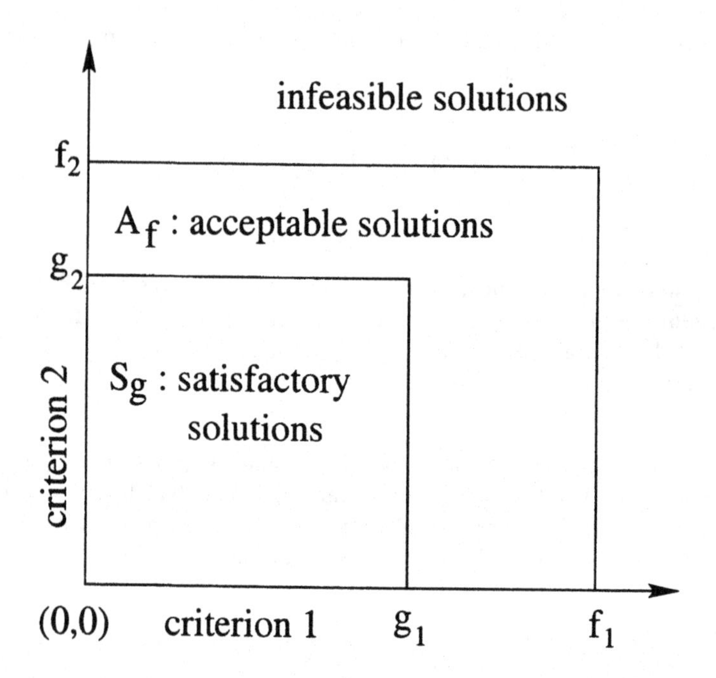

Figure 4.2 Sets of satisfactory and acceptable solutions for two dimensions

Definition 4.2 Let $x, y \in \Pi$. The relation x *is preferrable to* y *with respect to* $(g, f) \in G$, written $x \prec_{(g,f)} y$, is defined as follows: If x satisfy all goals, that is, $x \in S_g$, then

$$x \prec_{(g,f)} y \quad \Leftrightarrow \quad (x <_d y) \vee (y \notin S_g) \tag{4.2}$$

If x satisfies none of the goals, that is, $\forall\, i : c(x)_i > g_i$ then

$$x \prec_{(g,f)} y \quad \Leftrightarrow \quad (x <_d y) \vee [(x \in A_f) \wedge (y \notin A_f)] \tag{4.3}$$

Finally, x may satisfy some but not all goals.
Assume[1] that $(\forall\, i < k : c(x)_i \leq g_i) \wedge (\forall\, i \geq k : c(x)_i > g_i)$. Then

$$x \prec_{(g,f)} y \quad \Leftrightarrow \quad [(\forall\, i \geq k : c(x)_i \leq c(y)_i) \wedge (\exists\, i \geq k : c(x)_i < c(y)_i)] \tag{4.4}$$

$$\vee$$

$$[(x \in A_f) \wedge (y \notin A_f)] \tag{4.5}$$

[1]Since this can always be obtained by ordering the optimization criteria, full generality is preserved.

$$\vee$$

$$[(\forall\, i \geq k : c(x)_i = c(y)_i) \wedge \tag{4.6}$$

$$\{((\forall\, i < k : c(x)_i \leq c(y)_i) \wedge$$

$$(\exists\, i < k : c(x)_i < c(y)_i)) \tag{4.7}$$

$$\vee$$

$$(\exists\, i < k : c(y)_i > g_i)\}] \tag{4.8}$$

Notice that in Expression (4.4) only cost values of non-satisfactory dimensions are considered. In other words, when two solutions satisfy the same subset of goals, they are considered equal with respect to the satisfactory dimensions, regardless of their specific cost values in these dimensions. However, in the special case where two solutions are equal with respect to all non-satisfactory dimensions (see Expression (4.6)) their cost values in satisfactory dimensions determines their ordering (see Expression (4.7) and (4.8)).

The relation introduced in Definition 4.2 is an extension of the relation introduced in [64], adding the feasibility vector f and the notion of acceptable solutions, A_f. In [58] the relation is applied to control the search of an EA based algorithm for floorplanning, and the extension of the relation is observed to be of significant practical value. f restricts the EA based search to the region of practical interest by preventing the algorithm from wasting time exploring solutions which are non-dominated but in practice infeasible.

When it is clear which goal and feasibility vectors (g, f) are used, $\prec_{(g,f)}$ is written $\prec$. Furthermore, $\neg(x \prec y)$ is written $x \not\prec y$. The extended definition of $\prec$ is transitive, as is intuitively needed:

Theorem 4.1 $\forall\,(g, f) \in G, \forall\, x, y, z \in \Pi : x \prec y \wedge y \prec z \Rightarrow x \prec z$

Proof: As mentioned above, $\prec$ is a modification of the relation introduced in [64], denoted here by $\prec_p$. Specifically, comparing to [64] it can be seen that by construction:

$$\forall\, x, y \in \Pi : x \prec y \quad \Leftrightarrow \quad (x \prec_p y) \vee (x \in A_f \wedge y \notin A_f) \tag{4.9}$$

Given $(g, f) \in G$, $x, y, z \in \Pi$, and assume $x \prec y \wedge y \prec z$. Then

$$[(x \prec_p y) \vee (x \in A_f \wedge y \notin A_f)] \quad \wedge \quad [(y \prec_p z) \vee (y \in A_f \wedge z \notin A_f)] \tag{4.10}$$

For this to be true, truth values can be assigned to the clauses of the expression in a total of nine distinct ways. Four of these imply

$$y \notin A_f \wedge y \in A_f$$

and hence are impossible. Three others imply

$$x \prec_p y \wedge y \prec_p z.$$

Since $\prec_p$ is shown in [64] to be transitive, $x \prec z$ then follows from Expression (4.9). Two cases remain:

Case a) $x \in A_f \wedge y \notin A_f \wedge y \prec_p z$:
$y \notin A_f \Rightarrow y \notin S_f$. If y satisfies no goals, it follows from the definition of $y \prec_p z$ that $y <_d z$ and hence that $z \notin A_f$, which implies $x \prec z$ due to Expression (4.9). If y satisfies some but not all goals,

$$\exists k : (\forall i < k : c(y)_i \leq g_i) \wedge (\forall i \geq k : c(y)_i > g_i).$$

Then $y \notin A_f \Rightarrow \exists j \geq k : c(y)_j > f_j$. Either Expression (4.4) or (4.6) in the definition of $y \prec_p z$ is true. In either case,

$$c(y)_j \leq c(z)_j \Rightarrow c(z)_j > f_j \Rightarrow z \notin A_f \Rightarrow x \prec z.$$

Case b) $x \prec_p y \wedge y \in A_f \wedge z \notin A_f$:
Substituting x for y and y for z, it was shown in case a) that

$$x \notin A_f \wedge x \prec_p y \Rightarrow y \notin A_f.$$

By negation, $y \in A_f \Rightarrow x \in A_f \vee x \not\prec_p y$. Since $y \in A_f \wedge x \prec_p y$ then $x \in A_f$, from which $x \prec z$ follows due to Expression (4.9).

□

Using $\prec$, a given set of solutions can now be ranked: Define the mapping $r : \Pi \times 2^{\Pi} \mapsto N$ by

$$\forall y \in \Pi, \forall X \subseteq \Pi : r(y, X) = |\{x \in X | x \prec y\}|.$$

$r(y, X)$ is the rank of y with respect to X. Furthermore, let $X_0 = \{x \in X | r(x, X) = 0\} \subseteq X$. I.e., X_0 is the subset of best solutions in X with respect to $\prec$.

The relation $\prec$ have the following additional properties:

1. $\forall x \in \Pi : x \not\prec x$, i.e. $\prec$ is not reflexive. This follows from Expression (4.2) through (4.8).

2. From transitivity: $\forall x, y \in \Pi : x \prec y \wedge y \prec x \Rightarrow x \prec x$. But since $\prec$ is not reflexive, $\forall x, y \in \Pi : \neg(x \prec y \wedge y \prec x)$. Consequently, $\prec$ *is* antisymmetric (since the assumption of antisymmetry is always false), but since $\prec$ is not reflexive, it is not a partial ordering [26].

3. $\forall x, y \in \Pi : x \prec y \wedge x \notin S_g \Rightarrow y \notin S_g$. This follows from Expression (4.2): $y \in S_g \Rightarrow y \prec x$, contradicting $\neg(x \prec y \wedge y \prec x)$ above.

4. $\forall x, y \in \Pi : x \prec y \wedge x \notin A_f \Rightarrow y \notin A_f$. This follows similarly from Expression (4.2), (4.3) and (4.5): $y \in A_f \Rightarrow y \prec x$, contradicting $\neg(x \prec y \wedge y \prec x)$ above.

5. $\forall X \in \Pi : X \cap S_g \neq \emptyset \Rightarrow X_0 \subseteq S_g$. This follows from Expression (4.2): $y \notin S_g \Rightarrow \forall x \in X \cap S_g : x \prec y \Rightarrow r(y, X) > 0 \Rightarrow y \notin X_0$.

6. $\forall X \in \Pi : X \cap A_f \neq \emptyset \Rightarrow X_0 \subseteq A_f$. This follows from Expression (4.2), (4.3) and (4.5): $y \notin A_f \Rightarrow \forall x \in X \cap A_f : x \prec y \Rightarrow r(y, X) > 0 \Rightarrow y \notin X_0$.

7. In the special case of $g = (0, \ldots, 0)$ and $f = (\infty, \ldots, \infty)$ it can be seen from Expression (4.2) through (4.8) that $x \prec y$ is equivalent to $x <_d y$.

8. In the special case of $n = 1$, i.e. one-dimensional optimization, it can be seen from Expression (4.2) and (4.3) that $x \prec y$ is equivalent to $x < y$, regardless of $(g, f) \in G$.

4.3.2 Comparing Sets of Solutions

From one or more given sets of solutions the user will, sooner or later, select a single specific solution as the "best". For example, in the case of circuit design, a single layout will ultimately be selected for production. However, it is not known *how* the user makes the final choice; it depends on preferences which may never be explicitly expressed. But independently of the final selection method, if the "best" solution belongs to set X rather than set Y, then set X was more valuable to the user than Y and hence should intuitively be better than Y.

The basic idea of the proposed solution set quality measure is to model the final selection performed by the user by an explicit selection function which is parameterized to account for a wide range of possible preferences with respect to the relative importance of the optimization criteria considered. By systematically varying the parameters of the selection function, a class of functions corresponding to a wide range of possible user-preferences is obtained, and the quality of solution set X is then defined as the expected value of the selection function when selecting from X while traversing the class of selection functions.

Assume that the user who ultimately selects a single solution also defines (g, f) used by some algorithm to generate the set(s) from which the final selection is made. The model of the selection process should then be consistent with the preference relation $\prec_{(g,f)}$, assuming that the users actions are consistent. Obtaining this consistency will guide the construction of the set quality measure.

Before proceeding with a model of the selection process, a normalization procedure is needed. Different optimization criteria will have distinct units, for example mm^2, ns, % and $, and will have absolute values in very different ranges. A normalization is therefore performed initially in order to make comparisons in the multi-dimensional cost space without a priori introducing any bias towards a specific dimension.

Definition 4.3 Given $X \subseteq \Pi$ and $(g, f) \in G$. Let $c_{i,max} = \max_{x \in X}\{c(x)_i\}$, $c_{i,min} = \min_{x \in X}\{c(x)_i\}$ $i = 1, \ldots, n$, and assume[2] that $\forall\, i : c_{i,max} > c_{i,min}$. Furthermore, let

$$\Omega(X) = \{x \in \Pi | \, \forall\, i : c_{i,min} \leq c(x)_i \leq c_{i,max}\} \subseteq \Pi$$

and let $\epsilon > 0$ be a small, arbitrary constant. Then $\eta : \mathbf{R}^n_{+\infty} \mapsto [0; 1]^n$ is defined by

$$\forall\, i = 1, \ldots, n \;\; : \;\; \eta(t)_i = \begin{cases} 0 & \text{if } t_i < c_{i,min} \\ (1 - 2\epsilon)\frac{t_i - c_{i,min}}{c_{i,max} - c_{i,min}} + \epsilon & \text{if } c_{i,min} \leq t_i \leq c_{i,max} \\ 1 & \text{if } t_i > c_{i,max} \end{cases}$$

The normalized cost $\bar{c} : \Omega(X) \mapsto [\epsilon, 1 - \epsilon]^n$ with respect to $\Omega(X)$ is defined as $\bar{c}(x) = \eta(c(x))$. Similarly, the normalized goal vector $\bar{g}$ of g with respect to $\Omega(X)$ is defined as $\bar{g} = \eta(g)$.

By using $\bar{c}$ and the corresponding normalized goal and feasibility vectors $\bar{g}$ and $\bar{f} = \eta(f)$, the effects of different scalings of the cost dimensions are eliminated,

[2] If $c_{i,max} = c_{i,min}$ for some i, all solutions in X are equal with respect to the ith dimension, which can then be eliminated from the comparison, i.e. full generality is preserved.

allowing cost values to be compared while preserving the relative ordering defined by $\prec$.

The final selection performed by the user is modeled by the selection function s_w defined in Definition 4.4 below. The idea is that s_w should be as simple as possible, while still being consistent with $\prec$ in the sense stated in Theorem 4.2. s_w is essentially a weighted sum of all optimization criteria, which is probably the simplest, meaningful selection function. The scaling terms $w \cdot \bar{g}$ and $\sum_{i=1}^{n} w_i$ as well as the max-terms are minimal alterations of a weighted sum required to obtain the consistency with $\prec$. As mentioned earlier, since the user defines both $(g, f) \in G$ and hence $\prec_{(g,f)}$, it seems intuitively reasonable that the final selection performed using s_w should satisfy Theorem 4.2.

Definition 4.4 Given $X \subseteq \Pi$, $(g, f) \in G$, and a weight vector $w \in \mathbf{R}_+^n$. The *selection function* $s_w : \Omega(X) \mapsto \mathbf{R}_+$ is defined by

$$s_w(x) = \begin{cases} \sum_{i=1}^{n} w_i \bar{c}(x)_i & \text{if } x \in S_g \\ w \cdot \bar{g} + \sum_{i=1}^{n} w_i \max(\bar{c}(x)_i, \bar{g}_i) & \text{if } x \in A_f \setminus S_g \\ w \cdot \bar{g} + \sum_{i=1}^{n} w_i + \sum_{i=1}^{n} w_i \max(\bar{c}(x)_i, \bar{g}_i) & \text{if } x \notin A_f \end{cases}$$

where $\bar{c}$ and $\bar{g}$ are normalized with respect to $\Omega(X)$ according to Definition 4.3.

The notion of selection function consistency can now be formulated and proven:

Theorem 4.2 Given $s_w : \Omega(X) \mapsto \mathbf{R}_+$ according to Definition 4.4.

$$\forall\, x, y \in \Omega(X),\ \forall\, w \in \mathbf{R}_+^n : x \prec y \Rightarrow s_w(x) \leq s_w(y)$$

Proof: First observe from Definition 4.3 that each coordinate function of η is nondecreasing, i.e.

$$\forall\, i = 1, \ldots, n,\ \forall\, u, v \in \mathbf{R}_{+\infty}^n : u_i \leq v_i \quad \Rightarrow \quad \eta(u)_i \leq \eta(v)_i \qquad (4.11)$$

Consequently, $\forall\, x \in S_g \cap \Omega(X)$, $\forall\, w \in \mathbf{R}_+^n$:

$$c(x)_i \leq g_i \Rightarrow \bar{c}(x)_i \leq \bar{g}_i \Rightarrow \sum_{i=1}^{n} w_i \bar{c}(x)_i \leq w \cdot \bar{g}.$$

Since all terms in the definition of s_w are positive,

$$\forall\, x \in S_g \cap \Omega(X),\ \forall\, y \notin S_g \cap \Omega(X),\ \forall\, w \in \mathbf{R}_+^n : s_w(x) \leq s_w(y) \qquad (4.12)$$

Similarly, $\forall x \in \Omega(X) : \max(\bar{c}(x)_i, \bar{g}_i) \leq 1 \Rightarrow \sum_{i=1}^n w_i \max(\bar{c}(x)_i, \bar{g}_i) \leq \sum_{i=1}^n w_i$, and therefore, from the definition of s_w,

$$\forall x \in A_f \cap \Omega(X), \forall y \notin A_f \cap \Omega(X), \forall w \in \mathbf{R}_+^n : s_w(x) \leq s_w(y) \tag{4.13}$$

Given fixed $x, y \in \Omega(X)$, $w \in \mathbf{R}_+^n$ and assume $x \prec y$. First consider the case $x \in S_g$. If $y \notin S_g$, then $s_w(x) \leq s_w(y)$ follows from Expression (4.12). If $y \in S_g$, then from Expression (4.2) and (4.11), $x <_d y \Rightarrow \forall i : c(x)_i \leq c(y)_i \Rightarrow s_w(x) = \sum_{i=1}^n w_i \bar{c}(x)_i \leq \sum_{i=1}^n w_i \bar{c}(y)_i = s_w(y)$. Then consider the case $x \notin S_g$. If $x \in A_f \setminus S_g$ then $y \notin S_g$ according to Property 3 on page 31. On the other hand, if $y \notin A_f$ the result follows from Expression (4.13). Similarly, if $x \notin A_f$, then $y \notin A_f$ according to Property 4 on page 31. Hence, only two possibilities remain:

$$(x \in A_f \setminus S_g \wedge y \in A_f \setminus S_g) \quad \vee \quad (x \notin A_f \wedge y \notin A_f) \tag{4.14}$$

In either case, the same sub-expression of s_w is used when computing $s_w(x)$ and $s_w(y)$. Hence, it follows from Definition 4.4 that it is sufficient to show that

$$\sum_{i=1}^n w_i \max(\bar{c}(x)_i, \bar{g}_i) \quad \leq \quad \sum_{i=1}^n w_i \max(\bar{c}(y)_i, \bar{g}_i) \tag{4.15}$$

for all x and y satisfying Expression (4.14). Assuming Expression (4.14), consider the case when x does not satisfy any goal. Then from Expression (4.3), $x <_d y \Rightarrow \forall i : \bar{c}(x)_i \leq \bar{c}(y)_i$, from which Expression (4.15) follows. Finally, assume that x satisfies some but not all goals. As in Definition 4.2, assume without loss of generality that $(\forall i < k : c(x)_i \leq g_i) \wedge (\forall i \geq k : c(x)_i > g_i)$ for some k, $2 \leq k \leq n$. Since either Expression (4.4) or (4.6) holds,

$$\begin{aligned}
\sum_{i=1}^n w_i \max(\bar{c}(x)_i, \bar{g}_i) &= \sum_{i=1}^{k-1} w_i \max(\bar{c}(x)_i, \bar{g}_i) + \sum_{i=k}^n w_i \max(\bar{c}(x)_i, \bar{g}_i) \\
&\leq \sum_{i=1}^{k-1} w_i \bar{g}_i + \sum_{i=k}^n w_i \max(\bar{c}(x)_i, \bar{g}_i) \\
&\leq \sum_{i=1}^{k-1} w_i \max(\bar{c}(y)_i, \bar{g}_i) + \sum_{i=k}^n w_i \max(\bar{c}(y)_i, \bar{g}_i) \\
&= \sum_{i=1}^n w_i \max(\bar{c}(y)_i, \bar{g}_i)
\end{aligned}$$

□

Using s_w a solution quality measure $q(Y)$ of a set Y can now be defined. The idea is to vary the weight vector w used in s_w over a user-defined space W, and then define $q(Y)$ as the expected value of s_w over W when selecting solutions from Y. In other words, the quality of a set is the expected selection function value when varying the weights of the selection function corresponding to a wide range of possible preferences of the user. Notice that a smaller value of the quality measure q means a higher set quality.

Definition 4.5 Given $s_w : \Omega(X) \mapsto \mathbf{R}_+$ according to Definition 4.4 and a weight space $W \subseteq \mathbf{R}_+^n$. The *set quality measure* $q : 2^{\Omega(X)} \mapsto \mathbf{R}_+$ is defined by

$$\forall\ Y \subseteq \Omega(X)\ :\ q(Y) = \mathrm{E}(\min_{y \in Y}\{s_w(y)\})$$

where the expected value E is computed over all $w \in W$.

As an immediate consequence of Theorem 4.2, $q(Y) = q(Y_0)$, i.e. only solutions in Y_0 contributes to $q(Y)$, which is intuitively reasonable. Furthermore, the definition of q is independent of the weight space W. Hence, W can be discrete or continuous, and the weights can be independent or dependent. For example, W can be defined by assuming that w_i is normally distributed $N(\mu_i, \sigma_i)$, or uniform on $[0, m_i]$. Any available knowledge of the preferences used when making the final choice of a solution can thus be utilized when defining W. Due to the normalization performed according to Definition 4.3, a reasonable weight space can be defined without having any knowledge of the absolute cost values of each dimension. For example, allowing each w_i to take on any value in $[0, 1]$ is sufficient to capture all relative priorities of the criteria.

Theorems 4.3 and 4.4 below states two properties of the solution set quality measure, which are intuitively required. Theorem 4.3 states that by adding a solution to a set, the quality will be unchanged or will be improved. Theorem 4.4 states that when set Y is preferable to set Z in the obvious sense that the best solutions of $Y \cup Z$ all belongs to Y, then Y is at least as good as Z.

Theorem 4.3 Given $q : 2^{\Omega(X)} \mapsto \mathbf{R}_+$ according to Definition 4.5.

$$\forall Y \subseteq \Omega(X), \forall y \in \Omega(X) : q(Y \cup \{y\}) \leq q(Y)$$

Proof: Let $q : 2^{\Omega(X)} \mapsto \mathbf{R}_+$, $Y \subseteq \Omega(X)$ and $y \in \Omega(X)$ be given. Since the minimum value of any set is non-increasing as the set is expanded,

$$\forall w \in \mathbf{R}_+^n :\ \min_{x \in Y \cup \{y\}}\{s_w(x)\}\ \leq\ \min_{x \in Y}\{s_w(x)\}. \tag{4.16}$$

$q(Y \cup \{y\}) \leq q(Y)$ then follows from the properties of the expected value E, independently of the weight space W. □

Theorem 4.4 Given $q : 2^{\Omega(X)} \mapsto \mathbf{R}_+$ according to Definition 4.5.

$$\forall Y, Z \subseteq \Omega(X) : (Y \cup Z)_0 = Y_0 \Rightarrow q(Y) \leq q(Z)$$

Proof: Let $q : 2^{\Omega(X)} \mapsto \mathbf{R}_+$ and $Y, Z \subseteq \Omega(X)$ be given and assume that $(Y \cup Z)_0 = Y_0$. Similar to the proof of Theorem 4.3, it is sufficient to show that

$$\forall w \in \mathbf{R}_+^n : \min_{y \in Y}\{s_w(y)\} \leq \min_{z \in Z}\{s_w(z)\}. \tag{4.17}$$

For a fixed $w \in W$, let $z' \in Z$ be an element minimizing s_w, i.e.

$$s_w(z') \leq \min_{z \in Z}\{s_w(z)\}.$$

Since no solution in $Y \cap Z$ can cause a violation of Expression (4.17), assume without loss of generality that $Y \cap Z = \emptyset$. Then

$$(Y \cup Z)_0 = Y_0 \Rightarrow \forall z \in Z , \exists y \in Y_0 : y \prec z$$

and consequently, $\exists y' \in Y_0 : y' \prec z'$. From Theorem 4.2 $s_w(y') \leq s_w(z')$ from which Expression (4.17) follows since $\min_{y \in Y}\{s_w(y)\} \leq s_w(y')$. □

Intuitively it seems desirable that adding a solution y to X should be guaranteed to improve the set quality if $r(y, X) = 0$. Similarly, when set Y contains solutions dominating every solution in set Z, set Y should be strictly better than Z. In other words, the inequalities in Theorems 4.3 and 4.4 should ideally be strict, rather than allowing equality. This limitation is a consequence of the general shortcoming of the proposed set quality measure that non-convex points of the cost trade-off surface are not credited. More specifically, given y and X such that $r(y, X) = 0$, if $c(y)$ is non-convex relative to the cost of the solutions X_0, then $q(X \cup \{y\}) = q(X)$. The reason is that s_w is essentially a weighted sum, and the minimum of a weighted sum can never correspond to a non-convex point [64]. Therefore, y never minimizes s_w. The problem of crediting non-convex points is further discussed in the following.

4.3.3 Practical Issues of Set Comparison

Assume that two stochastic algorithms A and B have been executed m times each, generating solution sets $A_1, \ldots, A_m$ and $B_1, \ldots, B_m$, respectively. The practical computation of $q(A_1), \ldots, q(A_m), q(B_1), \ldots, q(B_m)$ relative to a goal and feasibility vector pair $(g, f) \in G$ is described in this section. When all set quality values are known, questions such as "Is A better than B?" can be answered by applying statistics on the set quality values. A and B can be based on any search strategy, and need not apply the same strategy. If the use of goal and feasibility values are considered inappropriate for the comparison, $g = (0, \ldots, 0)$ and $f = (\infty, \ldots, \infty)$ can be applied, which in effect eliminates the notions of goals and feasibility.

The solution quality values are determined by going through the following steps:

1. Normalize the cost dimensions according to Definition 4.3, using

$$X = \bigcup_{i=1}^{m} (A_i \cup B_i)$$

   and any small ϵ, for example $\epsilon = 10^{-4}$. Also normalize the goal vector. By construction, $\Omega(X)$ is now large enough for all remaining computations to be well-defined. The quality $q(V)$ of a new set V can be computed using the same normalization as long as $V \subseteq \Omega(X)$. However, if $V \not\subseteq \Omega(X)$, it is necessary to re-normalize all sets considered.

2. Define a weight space W. There are no restrictions on the definition of W, any available knowledge of the selection process can be incorporated (see Section 4.3.2).

3. Depending on the choice of W, an expression for the exact value of q may be obtainable and can then be used to accurately calculate q for each solution set. However, a simpler approach which applies to any definition of W is to compute an estimate $\hat{q}$ of q by sampling N points. I.e. for each set Y, estimate $q(Y)$ by

$$\hat{q}(Y) = \frac{1}{N} \sum_{k=1}^{N} \min_{y \in Y} \{s_{w(k)}(y)\}$$

   where each vector $w(k)$ is generated independently according to the chosen definition of W. Assuming a sufficiently large sample size N, this estimation method is statistically sound. Furthermore, Theorem 4.3 and 4.4 still holds when substituting q by $\hat{q}$, as a consequence of Expression (4.16) and (4.17), respectively.

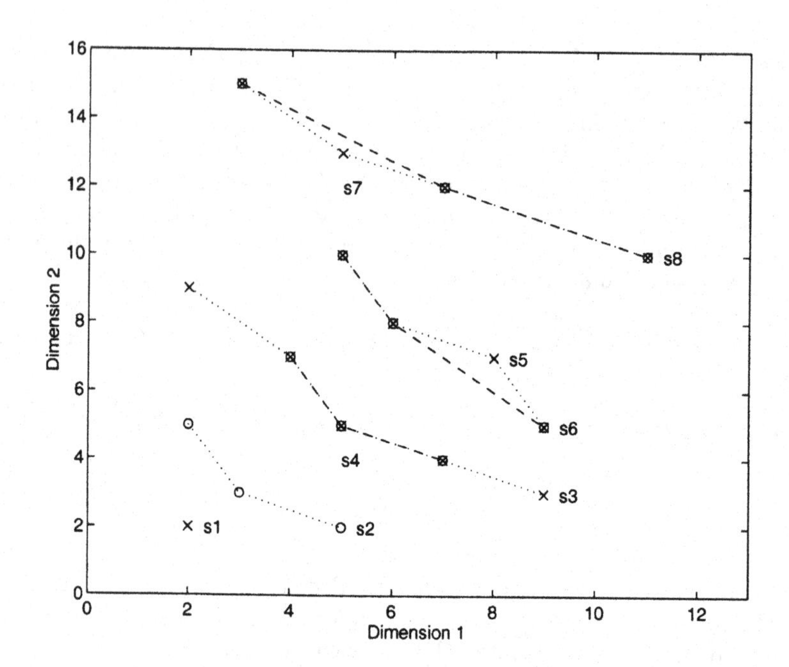

Figure 4.3 Constructed sets s1 through s8

4.3.4 Experimental Results

In this section the proposed quality measure is empirically validated using constructed sets. The measure is used in [58] to compare the performance of a EA for floorplanning with a random walk (see also Section 6.3.3).

Assume first that $n = 2$, $g = (0, 0)$ and $f = (\infty, \infty)$. Eight constructed sets s1 through s8 are plotted in Figure 4.3. Each set is indicated by either circles connected by a dotted line or crosses connected by a dashed line. s4 is a subset of s3, s6 a subset of s5 and s8 a subset of s7. The quality of each set is estimated as described above, using $W = [0, 1]^2$ and assuming that the weights are independent and uniform on $[0, 1]$. A sample size of $N = 10,000$ is used when computing $\hat{q}$.

When computing $\hat{q}$ a solution may never be sampled, i.e. never minimize s_w. This will happen if either

Set	s1	s2	s3	s4	s5	s6	s7	s8
Size	1	3	5	3	4	3	4	3
$\hat{q}$	0.0001	0.0768	0.2143	0.2655	0.4149	0.4149	0.5270	0.5291

Table 4.1 Estimated solution set qualities

1. the solution is non-convex (see Section 4.3.2) or
2. the weight w needed for the solution to minimize s_w was never generated during the computation of $\hat{q}$.

In the case of sets s1 through s8, all non-sampled points are non-convex.

As can be seen from Table 4.1, except for sets s5 and s6, the set quality decreases strictly with the set number, which seems intuitively reasonable. However, set s5 is not better than s6 because the solution with cost (8,7) is non-convex relative to s5. Solutions (7,4) and (4,7) of set s3 and (7,12) of s7 are similarly not sampled due to their non-convexity.

The effect of changing (g, f) to $g = (9, 6)$ and $f = (13, 16)$ is illustrated in Figure 4.4 and Table 4.2, using four new constructed sets s1 through s4. The boxes indicates S_g and A_f. Using these values of (g, f) has the effect of changing the relative quality of the sets from the order s2, s1, s4, s3 (decreasing quality) to the order s1, s2, s3, s4, as is intuitively desirable. In Table 4.2 $\hat{q}_1$ is the estimated qualities using $g = (9, 6)$ and $f = (13, 16)$. $\hat{q}_2$ is computed using $g = (0, 0)$ and $f = (\infty, \infty)$.

4.3.5 Conclusions

A preference relation and a solution set quality measure has been proposed. The preference relation allows cost vectors to be compared directly, without resorting to single-valued, aggregated cost functions. The set quality measure allows performance evaluation of set-generating algorithms. Both the preference relation and the set quality measure are independent of the search method used and therefore generally applicable. At the same time, the set quality measure is consistent with the preference relation used to drive EA based searches, and it appropriately incorporates the specification of goal and feasibility vectors.

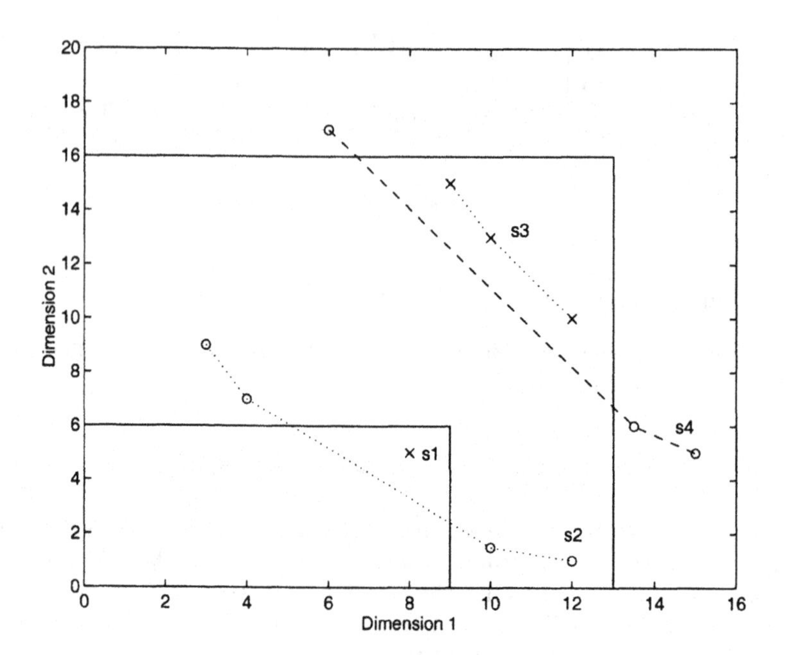

Figure 4.4 Constructed sets s1 through s4

Set	s1	s2	s3	s4
Size	1	4	3	3
$\hat{q}_1$	0.3347	0.8393	1.0328	1.9739
$\hat{q}_2$	0.3347	0.1807	0.6249	0.4963

Table 4.2 Estimated solution set qualities

The practical use of the measure has been illustrated on constructed sets and to a large extent the measure coincides with an intuitive notion of set quality. However, non-convex solutions are not appropriately accounted for, which is an important topic of future work. Other issues requiring further investigation includes sampling accuracy and the effect of increasing the number of optimization dimensions.

4.4 QUALITY VERSUS SPEED

So far when discussing algorithm performance only solution quality has been considered. But it should be emphasizied that this is not the only criterion. The runtime of an algorithm also has to be measured to evaluate its quality. In VLSI CAD the runtime of an algorithm is often simply measured in terms of CPU seconds on a workstation. Notice that this is not an ideal measure since the runtime largely depends on e.g. the coding style, the efficiency of the implementation, as well as the exact machine type (cache size, clock frequency, etc.), and not only the underlying algorithm. However, from a practical point of view, measuring absolute CPU time is still the best available measure of algorithm speed.

Runtime becomes even more important in VLSI CAD, since the problem instances are often very large and may consist of several thousand components, dependent on the application. When working with hard problems exact algorithms usually can not be applied. Instead, heuristic methods have to be used (see Chapter 1). Especially for EAs a fair runtime comparison is important, since due to the nature of the EA many solutions have to be evaluated, which often becomes very time consuming on large examples. These aspects will be discussed again in Section 6.5 after several applications have been presented.

4.5 SUMMARY

Methods for measuring performance have been discussed. When considering performance of an algorithm the quality of the solution obtained is probably the most important criterion. It has been discussed how different solutions can be compared without using an aggregated cost function and a theoretical background for such comparisons has been given.

In addition to solution quality the runtime consumption of an algorithm is also important. It is not surprising that an algorithm requiring "arbitrary" runtime will obtain good results. Therefore, the discussion of the runtime versus quality trade-off is essential and this issue should be explicitly addressed when proposing new algorithms, especially EAs.

PART II

PRACTICE

5

IMPLEMENTATION

5.1 INTRODUCTION

Several (recent) studies have shown that "pure" EAs are often not competitive with respect to runtime and quality to highly optimized problem specific heuristics (see e.g. [112, 73]). Thus, EA tools need efficient interfaces that allow to integrate existing techniques and that simplify software reuse.

In the following the basic underlying structure of an EA tool, called GAME (= Genetic Algorithm Managing Environment) [74], is presented. It has been developed over the past few years at the Albert-Ludwigs-University, Freiburg im Breisgau, Germany. GAME has in the meantime been successfully used in many applications in CAD of IC, i.e. logic synthesis, routing and testing. The main idea behind GAME is to provide a tool with a unified user interface that allows easy interaction with other VLSI CAD algorithms. By the discussion on GAME also an outline for the interested reader is given how to build his own EA environment.

In this chapter several aspects of the tool in comparison to standard implementations are discussed. Due to the requests of our applications, the modularity of the implementation is one of the key issues. This modularity is supported by a standardized *C++* interface used throughout the whole system. GAME is also the core software system of some of the applications discussed in Chapter 6. These applications demonstrate the flexibility of the tool and show how (very) different software tools can interact within the same environment.

The main idea in the following is not to describe a complete running *C++* environment. Instead many variants are presented that outline the flexibility of EAs. Which alternative is chosen depends on the specific applications where the EA should be applied.

5.2 EA TOOLS

In this section we briefly review "classical" EA tools and the different options they offer to the user. (For more details see [79, 34, 93, 111].) We repeat these (well-known) techniques so that the difference to the implementation described in the following becomes clear. All techniques discussed in this section are also supported by GAME.

5.2.1 Representation

Since different problems require different representations many EA tools support not only binary strings, but also multi-valued strings, graphs, sets, etc. The representation can be chosen with fixed or variable length. Often further restrictions can be overloaded with the representation, e.g. permutation problems require a string representation based on integer values with the constraint that each integer value only appears once.

5.2.2 Selection

The most frequently used selection schemes are roulette wheel selection and tournament selection. Furthermore, there exist several variations that are often problem specific, but they are in general not supported by each EA tool. GAME enables the possibility to let the designer integrate further problem specific selection schemes by himself.

5.2.3 Genetic Operators

The traditional operators are simple *crossover* and *mutation*. Many variations, like *2-point crossover* and *universal crossover*, are supported. Additionally, sequencing problems require crossover operators that guarantee that a valid permutation is obtained.

Example 5.1 *Partially Matched Crossover* (PMX) has been proposed as a genetic operator for the *Traveling Salesman Problem* [80]. PMX creates two children from two parents. The operator chooses two cut positions at random.

Notice that a simple exchange of the parts between the cut positions (as often applied to binary encoded EA problems) is not possible, since this would often produce invalid solutions. The operator works as follows to *validate* the children after the exchange:

PMX: Construct the children by choosing the part between the cut positions from one parent and preserve the position and order of as many variables as possible form the second parent.

For two elements p_1 and p_2 of length 3 the operator works as follows: Let $p_1 = 123$ and $p_2 = 231$ be the parents and let $i_1 = 1$ and $i_2 = 2$ be the two cut positions. The resulting children before the application of the *validation process* are $c_1' = 133$ and $c_2' = 221$. The validation procedure steps through the different elements between the cut positions (in the example only the second element) and restores the ordering. This results in the two valid children $c_1 = 132$ and $c_2 = 321$.

5.3 GAME: THE ENVIRONMENT

In this section the *Genetic Algorithm Managing Environment* (GAME) is described. First the basic underlying structure of the tool is outlined and some of its main features are highlighted. Then the hierarchical concept is discussed in more detail. It is shown how software reuse can easily be supported. The integration of heuristics in the EA tool is described. Finally, the interaction of different VLSI CAD tools within the GAME environment is outlined.

5.3.1 Basic Idea

Since the whole system is very complex, and the reader should not be confused with too many details, we start with an example of program interaction to demonstrate how different tools from different applications can communicate within the system. For this it is discuss how synthesis tools and testing tools can be used by the EA.

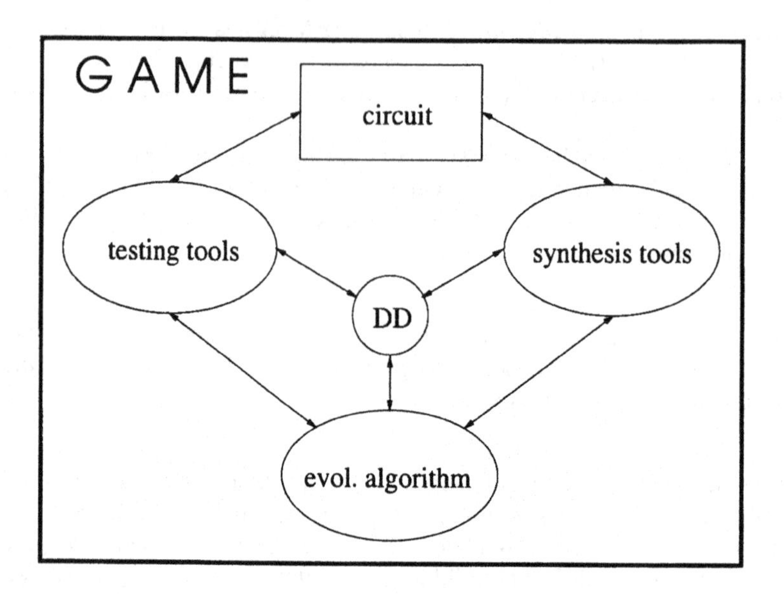

Figure 5.1 Sketch of the interaction scheme

Both kind of tools, i.e. synthesis tools and testing tools, have in common that they work on a circuit representation. Many tools have been developed over the past few years, e.g. the testing environment supports *fault simulators* and *test pattern generators*. The EA only needs communication with these tools, but a direct access to the circuit description is not needed. The same holds for the synthesis tools: The EA uses as a fitness measure only the information from the synthesis algorithms, but direct informations from the underlying circuit, i.e. from the object to be optimized, are not required.

A third component that should be mentioned here as an example are *Decision Diagrams* (DDs) [19, 39]. DDs are the state-of-the-art data structure in VLSI CAD and are used by testing tools and synthesis tools (see e.g. [147, 96, 159, 144]). Thus, they are separated from the individual tool set and can also directly interact with the EA system.

A sketch of the interaction scheme is given in Figure 5.1. It is important to notice the modular structure of the system: The EA has no direct access to the underlying circuit and can only access to it by the testing and synthesis tools.

5.3.2 A simple EA-Package

In the following the basic components of a EA package are discussed. The program code is given in a *C++*-like notation. The most basic element of an EA is the chromosome:

```
struct chromosome{
  int* code;
  long fitness;
  int  mutprob;
  int  age;
  ...
};
```

For simplicity it is assumed that the chromosome consists of an integer array. The element is then stored in variable `code`. Since to each chromosome a fitness is assigned the corresponding fitness value `fitness` is directly stored together with `code`. As an example also an individual mutation probability `mutprob` can be considered. Notice that as one potential of the EA each element in the population can be handled different. To guarantee that also new elements are used in a population, it is also possible to assign an age to each element, i.e. to consider how long this element already existed in the population. In many cases it has been shown to be advantageous to remove "old" elements, since otherwise they can dominate the population and can impede the EA to escape from local optima.

Next, the management of a single population is described. This will be used as a subclass in the following.

```
struct pop{
  long size;
  long fitsum;
  int* index;
  chromosome* chr_pntr;
};
```

`size` gives the size of the population, i.e. the current number of elements it contains. The sum of the fitness values of all elements in the population is stored in `fitsum`. This variable is often used to compute values needed for selection methods, like roulette wheel selection. In the array `chr_pntr` all chromosomes are stored. To allow an easy handling furthermore an index array index is introduced. By this array we can access to each element in chr_pntr directly with respect to some criteria, like e.g. fitness.

Finally, the class population is described. This class contains the information for handling the complete EA run. All available functions are declared as member functions in population.

```
class population{
  int no_of_pop;
  int size_of_chrom;
  int* size_of_pop;
  int crossover_prob;
  ...

  chrom best;
  ...

public:

  population();

  ~population();

  member_function1();
  member_function2();
  ...
}
```

In an EA not only a single population can be considered, but also optimization in parallel is often useful. Especially on a cluster of workstations this class concept allows an easy distribution of the program tasks. no_of_pop specifies the number of different populations. If the chromosome size is fixed the corresponding length is given in size_of_chromosome. The same holds for the size of the populations stored in size_of_pop. If a fixed crossover probability is used this is also stored in a separate variable crossover_prob. This variable can be manipulated by member functions. Similar variables are used for different types of mutation and variants of crossover. Since it is possible (due to the nature of the EA) that also the best element obtained so far is no longer present in the current population it makes sense to store it in a separate variable to make sure that it can be restored, if no better solution can be found. For this the best element computed is always stored in variable `best`.

For the member functions of class population some examples are given in the following. This is not to be seen as a complete list. Instead it should outline the principles of the underlying methods and should encourage the reader to develop his own functions that fit well in the corresponding application.

The first step in the EA is the initialization.

```
void  init_pop(param1, param2, ...){
  size_of_pop = param1;
  size_of_chrom = param2;
  ...
}
```

This can also be integrated directly in the constructor `population()` of the class population, but is more flexible in this way, since the initialization parameters can easily be changed. For this `population()` is mainly used for initializing the variables with default values. Dependent on the variables in class `population` the allocation of memory for arrays, etc. is done in `init_pop()`.

For a complete EA tool many subfunctions are often needed, e.g. random elements have to be created. Thus, functions like the followings must be available:

```
void  generate_random_chromosome();
void  generate_random_population();
```

(The names of the functions are self-explaining.)

Another important component is the availability of member functions that allow to get informations about the current status of the optimization process, e.g. the user wants to get the best result in the present population:

```
chrom get_best_chromosome_of_population();
```

Notice that this element might be different from the best element obtained so far, since the best element might be removed from the population dependent on the selection mechanism. Since in the end also the best result computed should be available (stored in variable best) a function is provided to output the best element:

```
chrom get_best_chromosome();
```

To analyze the most recent optimization step we also have to get informations about the lastly generated element:

```
chrom get_last_chromosome();
```

The newly generated chromosomes (generated by genetic operators, see below) are then inserted in the population:

```
void  insert_new_chromosome(chromosome*, int);
```

chrom denotes the chromosome to be inserted, while the second parameter specifies the position, i.e. the index, where to insert the element. Dependent on the selection principle e.g. the worst element is substituted by the newly generated one.

Before we start with the manipulation algorithms based on genetic operators, also some simple functions are helpful to handle chromosomes and get informations about them, like e.g. duplicating a chromosome and providing the computed fitness value:

```
chrom copy_chrom(chrom);
long fitness(chrom);
```

Based on the variables and functions presented so far the genetic operators can now easily be described. We start with the description of the implementation of a simple crossover operator:

```
chrom crossover(chromosome* ch1, chromosome* ch2){

  int i;
  int pos = random () % size_of_chrom;
  chrom child = new chrom;

  for (i = 0; i < pos; i++){
    child.code[i] = ch1->code[i];
  }
  for ( i = pos; i < size_of_chrom; i++){
    child.code[i] = ch2->code[i];
  }
  child.fitness = 0;
  child.age = 0 ;
  ...

  return child;
}
```

The newly created element `child` is generated by assigning the first entries (until position `pos`) from chromosome `ch1`, and the second part from chromosome `ch2`. Position `pos` is determined at random. The other variables in `child` (except array `code`) are initialized trivially. Dependent on the application also a direct call of the function computing the fitness value for the newly generated element can be integrated. In a similar way mutations and more complicated crossover operators can be defined:

```
chromosome* crossoverN(chromosome*,chromosome*);
chromosome* mutationI(chromosome*);
chromosome* mutationII(chromosome*);
chromosome* mutationN(chromosome*);
chromosome* inversion(chromosome*);
chromosome* PMX(chromosome*, chromosome*);
chromosome* OX(chromosome*, chromosome*);
chromosome* CX(chromosome*, chromosome*);
chromosome* Xmerge(chromosome*, chromosome*);
...
```

For a more detailed discussion on various genetic operators see [79, 34, 111]. All operators have in common that they have a very simple underlying structure and can be implemented easily in standard programming languages. An extension of the package outlined above is simple. In the following some more essential aspects beside the "pure" implementation are discussed that should further underline the advantages of a package that allows the interaction between different tools.

5.3.3 Software Reuse

Software reuse plays a more and more important role in the area of VLSI CAD. The tools become very complex, since many optimization criteria have to be considered during logic synthesis, e.g. area, delay and low power. Thus, it becomes impossible to develop efficient tools (within reasonable time bounds) without using existing source code.

All successful EA applications in the area of CAD of IC also intensively use existing methods (in most cases for the evaluation of the fitness function). In the system described here the EA makes use of well defined interfaces to communicate with this environment.

5.3.4 Hierarchical Concept

For most applications it is insufficient to make use of an EA package, where e.g. only string representations and "standard" crossover operators are supported. An example based on experiments that shows the necessity of alternative encoding schemes is given in the area of channel routing using evolutionary methods. Using a problem specific encoding in [78, 76] instead of a multi-valued string encoding improves the overall results with respect to runtime *and* quality. The reason is that the problem specific operators can work on these structures without time overhead. (See also Section 6.3.5.)

To overcome these problems GAME makes use of a hierarchical class concept as described above. Since the complete system is written in object-oriented *C++* language, it supports a modular structure and simplifies the understanding of the system.

Population class: The kernel of GAME is given by a class that handles the populations, the so-called *population manager*. The access to the present population and offsprings is managed, and also selection schemes and deletion schemes are included at this software level.

Data structure classes: The data structures that are available in GAME are specified in several classes. They determine the storage of the individuals. The most commonly used data structure in the EA environment is *class string*. The individuals are stored as multi-valued strings. Alternative classes for the representation of graphs, sets and several problem specific data structures for applications are available. All these classes can be used as template for the *population manager*. For each class at least one library of genetic operators is available.

Operator classes: For each data structure class a library supports genetic operators and/or problem specific operators. One library of *class string* contains the "standard" crossover operators, like *uniform crossover* and restrictions, i.e. *1-point crossover*, *2-point crossover*, and *mutations*. Additionally, sequencing problems are supported within this class: several permutation operators, like *PMX*, *ordered crossover*, etc. are available.

GAME enables the possibility to incorporate further data structures that can be handled by the population manager, like e.g. a tree structure.

5.3.5 Genetic Operators and Integration of Heuristics

One of the most important aspects that has to be considered in the area of VLSI CAD is the reuse of existing optimization techniques (as described above). Often (over the past few years) very efficient heuristics have been developed and the EA has to use them to be competitive with respect to quality *and* speed.

For this it must be easy to integrate "extern" heuristics in the EA operators. Due to the *C++* class concept the existing heuristics are available as member functions of e.g. the synthesis system. Thus, the EA can directly make use of them, similar to genetic operators.

5.3.6 Program Interaction

Many CAD of IC problems can not be seen isolated, e.g. during logic synthesis testing aspects should be considered to construct easily testable circuits (*design for testability*). Thus, the EAs have to access different software packages at the same time. The EA kernel in GAME handles several software packages in parallel, if the interface to the EA is given as described above.

5.4 APPLICATIONS

In the next chapter different application from each of the areas *logic synthesis*, *mapping*, and *testing* are presented. For each application the essential parts of integration of problem specific knowledge are highlighted. This turns out to be an important property in EA applications and that differ from standard implementations. By this it becomes clear that concepts like the ones used in GAME (or similar ones) are very important for successful VLSI CAD approaches.

5.5 SUMMARY

A *Genetic Algorithm Managing Environment* (GAME) has been presented that has been designed for applications in the area of VLSI CAD. In contrast to other environments, it is dedicated to integration of problem specific heuristics and thus towards software reuse. The basic communication structure has been outlined that enables a fast integration of EAs in the CAD environment, and vice versa.

6

APPLICATIONS OF EAS

6.1 INTRODUCTION

The first EA based algorithms for VLSI CAD that were of practical interest were presented in the late 80s and early 90s. In this chapter different CAD of IC application are considered, i.e. applications from logic synthesis, mapping, testing and EA based solutions are given. Several selected papers are reviewed, then extensions of these approaches are outlined. For each area, the detailed problem definition is given and the main characteristics of the algorithms are presented. Especially the performance evaluation methods are considered, since they are crucial for an EA approach in order to make an impact. The selection of papers is not complete in the sense that all approaches presented so far are considered. Instead alternative techniques give insight in the minimization principles of EAs.

6.2 LOGIC SYNTHESIS

The main idea in logic synthesis is to determine a "good" representation of a Boolean function with respect to a given design style. "Good" in this context means with respect to given optimization criteria, like e.g. area, delay, low power. First, two-level AND/EXOR forms are considered and an approach for 2-level minimization is studied in detail [6, 40]. Then general multi-level synthesis is considered.

6.2.1 AND/EXOR Minimization

One technology often used are *Programmable Logic Arrays* (PLAs), i.e. two-level circuits. Whereas powerful minimization tools for AND/OR based synthesis have been developed [32, 31], there is a lack of adequate tools for minimization of AND/EXOR circuits with a large number of inputs.

In contrast to AND/OR minimization in AND/EXOR minimization several subclasses are of interest [139]. The most general 2-level form is the *EXOR Sum-Of-Product expression* (ESOP). Several tools for minimization have been presented, but the exact algorithms can only be used for functions with a small number of variables (see e.g. [124, 153, 61]). For an overview and a comparison between the heuristic approaches which often try to apply methods known from an "ESPRESSO-like" AND/OR minimization see [139, 94].

In the last few years synthesis based on AND/EXOR realizations has gained more and more interest [139], because AND/EXOR realizations are very efficient for large classes of circuits, e.g. arithmetic circuits, error correcting circuits and circuits for tele-communication [140, 142, 139]. For these classes, the EXOR circuits derived from ESOPs, often need fewer gates for the representation of a Boolean function and drastically reduce the number of signals in the resulting network. Additionally, EXOR based circuits have good testability properties [129, 137, 48] - at least if they are restricted to specific subclasses of AND/EXOR forms - and thus are well suited for design for testability. The circuits can easily be mapped to FPGAs [17, 117], where EXOR gates, e.g. in table-lookup FPGAs, do not cause any extra cost in terms of chip area.

In the following *Fixed Polarity Reed-Muller expressions* (FPRMs), a restricted class of ESOPs, are considered. This problem is chosen due to several reasons:

- The problem of FPRM minimization is known for long [131] and in the meantime is well understood. Many methods to solve it have been proposed [136, 150, 90, 49].
- The problem definition is very simple (and for this is a good example to get started also for the "non-expert"):

  In FPRMs each variable may appear in complemented or uncomplemented form, but not both. The problem to be solved is to determine a good polarity for each variable, since the size of an FPRM may vary from linear to exponential (in the number of variables) depending on the chosen polarity (see below).

- FPRMs have "nice" properties, e.g. circuits resulting from FPRMs are easily testability [136, 48]. FPRMs are also used in logic synthesis [151]. Recently, it has been shown that FPRMs are also a good starting point for constructing easily testable multi-level circuits [152].

Fixed Polarity Reed-Muller Expression

An *FPRM* is an EXOR of AND product terms, where each variable only appears in complemented or uncomplemented form, but not in both. FPRMs are a canonical representation of Boolean functions $f : \mathbf{B}^n \to \mathbf{B}$, if the polarity for each variable is fixed. The choice of the polarity largely influences the size of the resulting FPRM, as is shown by the following example from [138]:

Example 6.1 Let $f : \mathbf{B}^n \to \mathbf{B}$, given by $f = \overline{x}_1\overline{x}_2..\overline{x}_n$. Thus, only one term is needed, if all variables are complemented. If all variables are uncomplemented the resulting expression consists of 2^n terms, i.e. $f = 1 \oplus x_1 \oplus x_2 \oplus ... \oplus x_1 x_2..x_n$.

For the representation of Boolean functions a multi-level data structure, called *Ordered Functional Decision Diagram* (OFDD), as defined in [49] is used:

A *Functional Decision Diagram* (FDD) is a graph based representation of a Boolean function $f : \mathbf{B}^n \to \mathbf{B}$ over the variable set X_n (similar to BDDs), where one of the following two decompositions is carried out in each node:

$$f = f_i^0 \oplus x_i f_i^2 \quad \textit{positive Davio (pD)} \tag{6.1}$$

$$f = f_i^1 \oplus \overline{x}_i f_i^2 \quad \textit{negative Davio (nD)} \tag{6.2}$$

(f_i^0 denotes the *cofactor* of f with respect to $x_i = 0$, f_i^1 denotes the cofactor for $x_i = 1$ and f_i^2 is defined as $f_i^2 := f_i^0 \oplus f_i^1$, $\oplus$ being the EXOR operation.)

If the underlying graph is *ordered*, i.e. if the variables appear in the same order along each path, the corresponding FDD is called an OFDD. Decomposition types are associated to the variables in X_n with the help of a *Decomposition Type List* (DTL) $d := (d_1, \ldots, d_n)$ where $d_i \in \{pD, nD\}$, i.e. for each variable one fixed decomposition is chosen.

On OFDDs reductions can be defined. From [49] it is known that reductions can be used to define canonical representations for OFDDs.

An OFDD for f directly corresponds to an FPRM of f with fixed polarity. Furthermore, OFDDs allow efficient synthesis operations, at least with respect to this application. An OFDD with a given polarity p is transformed to an OFDD with polarity p' by performing EXOR operations. Each EXOR operation on an OFDD G of size $|G|$ has runtime $O(|G|^2)$, and thus can be performed efficiently. Using the OFDD it is possible to determine very fast the number of terms of the corresponding FPRM and to change the polarity of the FPRM by manipulation of the OFDD. (For more details see [49].)

It can even be proved that OFDDs are a "good" data structure for FPRM minimization in the sense that OFDDs are never much larger than FPRMs, if they represent the same function. But in contrast the OFDD can be much smaller (even exponentially smaller) than the FPRM [38].

We now consider the following problem, that will be solved using EAs:

> *How can a polarity for a given Boolean function f be determined such that the number of terms in the corresponding FPRM is minimized?*

Evolutionary Algorithm

In this section the EA is described that is applied to the problem given above.

Representation

Often in applications of EAs the encoding problem is one of the most difficult. In FPRM minimization this is very easy, since the problem of FPRM minimization can directly be expressed in two-valued logic. For each variable a polarity must be chosen. Thus, each element of the population corresponds to an n-dimensional binary vector, where n denotes the number of variables. A population is a set of vectors. Using this binary encoding each string represents a valid solution. For the storage of the population a simple array structure can be used.

Objective Function and Selection

As an objective function that measures the fitness of each element the number of terms in the FPRM corresponding to the chosen polarity is used. This function has to be minimized to find a small two-level representation of the function. The number of terms is determined using OFDDs (as described above).

The selection is performed by ***roulette wheel selection***. Additionally, ***elitarism*** is used. The size of the population $|\mathcal{P}|$ is adapted dynamically using problem specific measures. (The choice of the population size will be explained in more detail later.)

Initialization

At the beginning of each EA run an initial population is randomly generated and the fitness is assigned to each element.

As outlined in Chapter 2 it is often helpful to combine EAs with problem specific heuristics, i.e. to use HEAs. For FPRM minimization the HEA makes use of the following (simple) ***greedy heuristic*** from [49]:

Greedy Heuristic (GRE): Start with an OFDD and change the polarity of variable x_1. Choose the best polarity for x_1 and go to the next variable (if it exists) and perform the same optimization. The heuristic stops, if all variables have been considered exactly once.

The initial population is further optimized by applying greedy algorithm GRE to i elements randomly chosen ($i \in \{0, .., pop\}$), where *pop* denotes the size of the population. (If $i = 0$ we have a "pure" EA, i.e. an EA without application of heuristics.) The application of GRE to the initial population guarantees that the starting points are not too bad, thus the convergence is speeded up. The heuristic GRE itself is time consuming, but forces a faster convergence. Since we are interested in both aspects, i.e. good results and fast runtimes, GRE is only applied at the beginning and at the end to the best element for a final optimization to avoid local minima.

The convergence behaviour of the HEA largely depends on the parameter i, i.e. the number of elements that are optimized by GRE in the initial population. Before the whole algorithm is described in detail this influence will be demonstrated by the considering benchmark *in2* from [160]:

Example 6.2 The HEA has been executed 3 times with varying choices for parameter i:

1. $i = 0$
2. $i = |\mathcal{P}|/4$
3. $i = |\mathcal{P}|$

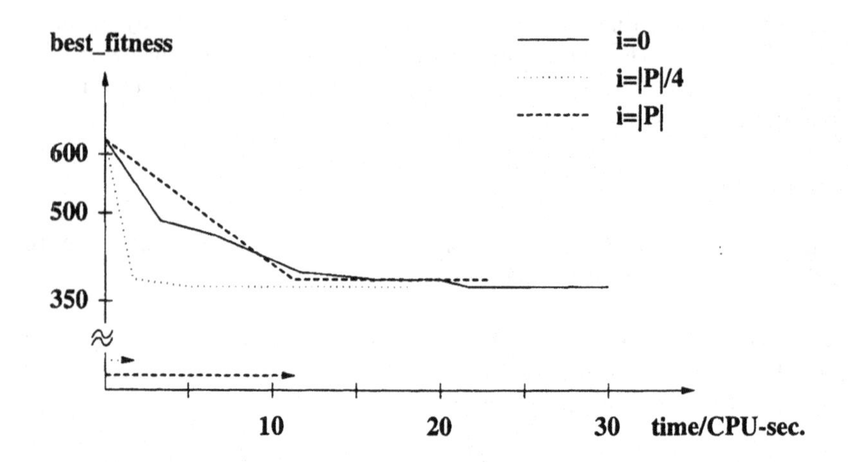

Figure 6.1 Influence of GRE for benchmark *in2*

The behaviour of the different EAs with respect to runtime and quality are given in Figure 6.1. The lines in the upper part show the number of terms, while the arrows in the lower part give the corresponding runtimes.

As can be seen the quality of the HEA that does not use GRE at all (solid line) is the best, i.e. 355 terms. But it also takes the largest runtime measured in CPU seconds: more than three times longer than the EA with $i = |\mathcal{P}|/4$. If only a quarter of the initial population is optimized by GRE (dotted line) the runtime is shortened tremendously and the result does not become worse. If the whole population is optimized at the beginning of the EA run (dashed line) the quality is decreased and the runtimes are not as good as for HEA using $i = |\mathcal{P}|/4$, because the time consuming heuristic GRE is applied too frequently. All in all, it turns out that GRE should be incorporated to speed up the convergence, but not too frequently, since otherwise the EA directly goes to a local minimum and it is not able to escape from that one any more.

This first example has shown that the parameters are very important in an EA to obtain a good trade-off between runtime and quality of the result.

Operators

In this first application only the operators introduced in Section 2.3.5, i.e. reproduction, (2-time) crossover, (2-time) mutation and mutation with neighbour,

```
hybrid_evolutionary_algorithm (problem instance, i):
    generate_random_population ;
    calculate_fitness ;
    if (HEA) optimize_elements_with_GRE (i)
    do
        apply_operators_with_corresponding_probabilities ;
        calculate_fitness ;
        update_population ;
    while (not terminal case) ;
    if (HEA) apply_GRE_to_best_element
    return best_element ;
```

Figure 6.2 Sketch of HEA

are used. Obviously, all genetic operators only generate valid solutions, if they are applied to binary strings, due to the encoding method used.

Algorithm

Using the operators introduced above the HEA works as follows:

1. Initially a random population of binary finite strings is generated and i elements are optimized by the greedy heuristic GRE.
2. The better half of the population is copied in each iteration without modification. Then the genetic operators *reproduction* and *crossover* are applied to another $\frac{|\mathcal{P}|}{2}$ elements. The newly created elements are then mutated by one of the three mutation operators with a given probability.
3. The algorithm stops if no improvement is obtained for $60 \cdot log(best\_fitness)$ iterations, where $best\_fitness$ denotes the fitness of the best element in the population. Finally, if $i > 0$, the greedy algorithm is applied to the best element.

A sketch of the algorithm is given in Figure 6.2.

Parameter Settings

The size of the population is adapted dynamically. The initial population has size half of the number of variables of the considered Boolean function. After initialization the population size depends on the average fitness of the population. If the average fitness is smaller than 100 the size is decreased to constant 5. Experimental results have shown that by this convergence and runtime are speeded up without loss of quality. For all benchmarks the same parameter set is used, i.e. the greedy heuristic is applied to $\frac{|\mathcal{P}|}{4}$ elements.

The genetic operators are iteratively called corresponding to their probabilities:

1. *Reproduction* is performed with a probability of 20%.
2. *Crossover* and *2-time crossover* are performed with a probability of 80%.
3. *Mutation*, *2-time mutation* and *mutation with neighbour* are carried out on the newly generated elements with a probability of 15%.

Experimental Results

In this section experimental results are given that were carried out on the package presented in [36] on a *Sun Sparc 10*. All times given are measured in CPU seconds.

The algorithm is applied to several benchmarks from LGSynth91 [160]. In [49] OFDD based heuristics have been compared to several other approaches. There it has been shown that with this method the best results were obtained. Thus, in the following we first restrict ourselves to a comparison with [49] with respect to runtime and quality of the results.

In a first series of experiments the HEA is applied to "small" benchmarks with up to 15 variables. For these functions the optimal solution with respect to the considered objective function has been determined in [49] by complete enumeration. The results are given in Table 6.1. *in* (*out*) denotes the number of inputs (outputs). Column *exact* gives the results of the exact method of [49] and HEA denotes the approach described above. *time* gives the runtime of the corresponding algorithm. *terms* denotes the number of terms in the resulting FPRM. The HEA always determined the minimal solution and additionally is much faster than the exact method, e.g. the HEA is more than 530 times faster for benchmark *gary*.

name	*in*	*out*	exact		HEA	
			terms	time	terms	time
add6	12	7	132	130.5	132	8.9
gary	15	11	349	9677.2	349	18.8
m181	15	9	67	436.4	67	3.2
tial	14	8	3683	4361.1	3683	101.2

Table 6.1 Exactly minimized FPRMs for small benchmarks

name	*in*	*out*	heuristic		HEA	
			terms	time	terms	time
apex7	48	37	604	5.4	515	157.8
bc0	26	11	1223	2.1	1117	128.0
bcb	26	39	1892	6.5	1892	455.0
chkn	29	7	900	5.5	900	83.2
cps	24	109	293	3.1	291	149.3
duke2	22	29	278	2.0	255	78.9
ibm	48	17	1181	7.3	1181	222.6
in2	19	10	359	0.6	355	18.7
in7	26	10	73	0.3	72	9.3
mish	94	43	54	3.6	53	127.1
s1196	32	32	7951	11.2	7012	793.6
vg2	25	8	5927	14.3	5290	133.0
x6dn	39	5	546	4.3	536	108.1

Table 6.2 Minimized FPRMs for large benchmarks

In a next series of experiments the HEA has been applied to larger benchmarks with up to 90 input variables. We first compare the results with the most powerful (and most time consuming) heuristic from [49]. The results are presented in Table 6.2. The HEA always determines the same or better results than the heuristic approach in [49]. Especially, for large functions the results of the HEA are much better than in the heuristic approach (see e.g. *apex7*), but the HEA is more time consuming.

name	*in*	*out*	exact		HEA	
			terms	time	terms	time
b9	16	5	105	7.8	105	7.8
bc0	26	11	1117	43.0	1117	128.0
cps	24	109	291	4.2	291	149.3
duke 2	22	29	255	30.0	255	78.9
in2	19	10	355	2.1	355	18.7
in7	26	10	72	3.8	72	9.3
mish	94	43	53	920.0	53	127.1
vg2	25	8	5290	3968.0	5290	133.0

Table 6.3 Exactly minimized FPRMs for large benchmarks

With a recently presented approach for some of the larger benchmarks also the optimal result can be determined [141][1]. To give an impression on the quality of the HEA, the results for some larger benchmarks are compared to [141] where the exact method could compute the result. As can easily be seen in Table 6.3 — column *exact* denotes the exact results from [141] — for these benchmarks the HEA always obtained the minimal number of terms. The runtimes for [141] are given in CPU seconds on an HP 715 with 256 MByte of main memory. In the case of "simple" functions, i.e. functions with simple decision diagram representation, the exact method even outperformed the HEA. On the other hand the HEA is still applicable where the exact method fails.

In addition to Example 6.2 finally the convergence behaviour of the HEA is discussed. This gives some more insight on the influence of the parameters:

Example 6.3 This time benchmark *vg2* is considered and two parameters are varied, i.e. the parameter i and the population size. The results are given in Table 6.4. i in the top row denotes the number of applications of the greedy heuristic after initialization. $|\mathcal{P}|$ denotes the size of the population. The upper row for each population size gives the resulting number of terms and the lower

[1]The method in [141] is based on EXOR ternary decision diagrams. The exact method can be applied as long as the corresponding decision diagrams can be constructed. Notice that in this method there is no direct correspondence between the number of inputs of a benchmark and the memory needed for the construction as a decision diagram, i.e. "simple" functions with large number of inputs do often cause no problems while small functions may be infeasible. This property also holds for the data structure OFDD that is used by the HEA, but OFDDs are often much smaller than EXOR ternary decision diagrams.

vg2	$i=0$	$i=\frac{\lvert\mathcal{P}\rvert}{4}$	$i=\frac{\lvert\mathcal{P}\rvert}{2}$	$i=\lvert\mathcal{P}\rvert$
$\lvert\mathcal{P}\rvert=1$	8682	8682	8682	8682
	8.6	10.0	10.0	10.0
$\lvert\mathcal{P}\rvert=5$	5519	5392	5392	5392
	68.2	70.5	99.5	53.1
$\lvert\mathcal{P}\rvert=12$	5290	5290	5316	5290
	207.2	133.0	156.2	146.8
$\lvert\mathcal{P}\rvert=25$	5290	5290	5290	5342
	212.4	301.8	263.6	210.2

Table 6.4 Results for benchmark *vg2*

row gives the runtime measured in CPU seconds. As can easily be seen good results are not obtained, if the population size is chosen too small. In contrast a too large population size wastes resources without any gain. The "pure" EA, i.e. the EA without application of the greedy heuristic, performs not very good, since the starting points are too bad. This avoids a fast convergence.

Conclusions

An OFDD based method to minimize FPRMs using HEAs was presented. The HEA focused on minimization of large functions and showed that the results improve significantly on the best heuristics known before.

This first (simple) application showed several features that are also important in the following:

- The genotype is a binary string. Its length is given by the number of variables of the Boolean function to be minimized, i.e. each bit in the string represents the polarity of the corresponding variable. The encoding was chosen in such a way that only valid elements occurred.

- The EA only used standard operators, like mutation and crossover, and can be combined with a *greedy heuristic*. This heuristic in general guarantees a fast convergence, but often reduces the quality of the result, if it is applied too frequently. Thus, the designer can choose between fast runtime and "poor" solutions or long runtime and "good" solutions.

- The EA has been tested against the best known heuristics. It was often able to further improve the size of the representation especially on large functions with more than 25 variables. Independently of the approach presented above a different method using decision diagrams has been presented [141], i.e. EXOR ternary decision diagrams have been used. Using this technique exact solutions for larger FPRMs have been computed. The exact algorithm confirmed all of the results: The EA never failed to obtain the exact solution. On the other hand where the exact approach failed the runtimes of the EA are much larger than the approaches using the greedy heuristic only. The combination of the EA based method with the greedy strategy allows the designer to explicitly control the trade-off between runtime and quality of the result.

The EA approach for FPRM minimization has been extended in [45] to deal with more general classes of AND/EXOR expressions. There similar results have been obtained.

In [35] an alternative approach for AND/EXOR minimization has been tried: An EA has been used, but no problem specific knowledge, neither in form of the operators nor by any heuristic, has been incorporated. The results obtained confirm the observations above, since the runtimes of the EA are long, but no better results than already existing heuristics and/or exact methods could be obtained. This underlines the need to incorporate problem specific knowledge in the design of EAs.

6.2.2 Multi-Level Minimization

Some Boolean functions do not have efficient two-level representations. For this, several heuristic minimization algorithms have been developed. Most of them are based on iterative application of network transformations. The order in which the transformations are applied largely influences the size of the resulting circuit. The sequences of transformations are often called *synthesis scenarios* or *scripts*. In [101] an EA was presented that determines a good script for given benchmark circuits. The scripts are specified as *general context-free grammars*. These grammars are translated into *directed graphs*. The genetic operators mutation and crossover are defined on the resulting graphs. It has been shown that the EA could further improve standard scenarios by 8% on average. On the other hand the runtimes are very large [101]: "Twenty-five workstations were employed for the evaluation of the scenarios. After 115 generations 37373 synthesis jobs were performed consuming a total of 2300 CPU hours (normalized to a 62 MIPS IBM RS/6000)."

An other approach based on *Decision Diagrams* (DDs) has been first presented in [41]. DDs are a graph based representation of Boolean functions (similar to Boolean networks), with some restriction on the decomposition that is carried out in each node and the structure of the circuit. (FDDs as introduced in the previous section are one example of a DD.) For a DD a good ordering of the input variables has to be determined. The size of DDs is very sensitive to the chosen variable ordering. Various methods (including greedy algorithms, dynamic optimization, simulated annealing, and genetic algorithms) have been presented (see e.g. [11, 122, 43]). In it simplest form the synthesis scenario reduces to finding a permutation of the variables (like in the case of the *traveling salesman problem*). The approach in [41] works with *Binary Decision Diagrams* (BDDs) and could determine the exact results for each function considered, but in much less time than the exact algorithm [66, 86]. Furthermore, the EA could also be applied to larger examples where the exact algorithm failed due to overly large runtimes and space requirements. In [44] this method has been applied to a more general type of DD, i.e. the *Kronecker Functional Decision Diagram* (KFDD). Here in addition to the ordering of the variables also a good decomposition has to be determined. Experiments demonstrated that in contrast to the classical greedy algorithm the EA could improve the result by up to a factor of two. A further extension in [42] also considered the testability of the resulting network as an optimization criterion. It was shown that the highly optimized networks (with respect to circuit size) often have poor testability properties. In contrast, if only testability has been considered as the optimization goal the circuits became very large. The approach from [42] allows the designer to choose (by setting parameters) between the two goals and the EA is used to find the best possible compromise. Summarizing all EA based DD approaches it can be stated that they in general outperform the previously published DD heuristics with respect to quality, but often are much slower than the greedy heuristics. Using more clever EA methods recent work [73, 46] has demonstrated that the runtime of the approach of [41] can significantly be reduced. This has been obtained by a combination of dynamic variable ordering and EAs. By this the methods are also applicable to functions with more than 200 variables. The algorithms become competitive not only with respect to quality (as most EAs), but also with respect to speed.

The EA approaches to logic synthesis described so far do not manipulate the netlist directly. Instead they use a command oriented language (script) or a restricted network type, like DDs. Recently, a new approach has been presented in [120]. The EA starts from a two-level form and optimizes the expression by transformation rules. The genes are given by the initial representation. For small circuits the approach obtains results comparable to SIS, but no runtimes are reported.

6.3 MAPPING

We now focus on the next step in the CDP: the mapping phase. Depending on the underlying technology different problem formulations are considered. We first outline an EA for technology mapping. Then partitioning, floorplanning, placement and global routing for ICs are considered. Finally, an approach for detailed routing is discussed.

6.3.1 Technology Mapping

First, technology mapping for FPGAs based on CLBs is considered. (For more details see [17, 117].) Each CLB can implement any Boolean function of k inputs, where k depends on the chosen library. Thus the problem to be solved is to find a CLB cover for a Boolean network such that the number of CLBs is minimized. In [92] GAFPGA, an EA based minimizer, has been presented. Each element in the population is represented by two strings of length N, where N denotes the number of gates in the network to be mapped. The first string stores the segmentation point placement in binary form and the second string represents the decomposition state, since gates with more than k inputs must be decomposed. Mutation and crossover are standard operators, but after each application a *validation procedure* is executed, since the genetic operators may create infeasible solutions. GAFPGA uses a very small population, i.e. only 4 elements, to reduce the runtime. The quality of the algorithm is measured on a set of benchmark circuits in comparison to several other state-of-the-art tools. The resulting circuits are smaller in most cases. A direct comparison of runtime is not given.

6.3.2 EA Approaches to Physical Design

Before the EA based algorithms to physical design are described in detail a brief overview is given: In some IC design styles, blocks consisting of transistors are positioned in the two-dimensional surface of the chip, and then interconnected. The positions are determined by first assigning each block to a specific region of the surface, which is called *partitioning*, and then determining the shape and position for all the blocks within each region, which is called *floorplanning*. A popular partitioning strategy is to recursively partition the blocks into two disjoint sets, and an EA for this, called GRCA, is presented in [20]. Here the problem is formulated in terms of a hypergraph $G = (V, E)$:

- Each vertex corresponds to a block and each hyperedge to a *net*.
- A net is a set of points, called *pins*, belonging to distinct blocks, which are to be connected, i.e. each hyperedge specifies a set of blocks.
- The bi-partitioning problem is to divide V into two disjoint sets A and B, such that a weighted sum of nets with pins in both A and B are minimized and such that A and B are approximately of the same size.

In GRCA, a genotype is a binary string of length equal to the number of blocks, where the ith bit is a one if and only if the ith block belongs to A. GRCA is a steady-state EA, and after each crossover and mutation, an attempt to improve the offspring is performed using a fast variant of the *Fiduccia-Mattheyses* (FM) partitioning heuristic. Before the EA itself is executed, the ordering of the genes are determined by a pre-processing step called *Weighted-DFS Reordering* (WDFR). In essence, the ordering is determined by a depth-first traversal of a graph constructed from the hypergraph G. The performance of GRCA is evaluated by comparison to two other partitioning approaches, using benchmark data sets. Averaging over all examples, GRCA produces better results than the other approaches while using a similar amount of runtime for the smallest graphs, and significant less runtime for the largest graphs. While the effect of WDFR on performance is inconclusive, it is noted that GRCA is not competitive without the use of the FM heuristic for local optimization.

Following partitioning, the blocks within each region are considered, one region at a time. For each block, a shape (aspect ratio), orientation (eight possibilities) and absolute position is determined. This problem is known as *floorplanning*. In the special case where the aspect ratio of all blocks are fixed, that is, the shape of each block is given, the problem is known as *placement*.

An EA based floorplanning approach is presented in [57, 58]. This floorplanner, called *Explorer*, minimizes layout area, aspect ratio deviation from a target value, maximum path delay and routing congestion. The key unique feature of Explorer is that it supports interactive design space exploration in the four-dimensional cost space. Rather than applying a scalar-valued, aggregated cost function, as virtually any other existing floorplanning tool, Explorer uses a vector-valued cost function to drive the search. Cost vectors are compared directly, without ever expressing solution cost by a single scalar value. The main purpose of this strategy is to avoid the traditional use of cost bounds and relative cost weights, since these quantities inherently introduce many practical problems, as discussed in Section 4.3. Explorer allows explicit exploration of alternative trade-offs of the four cost dimensions, and upon termination it outputs a set of alternative trade-offs. This is in contrast to traditional approaches, which output a single solution representing a specific cost compromise. The design space exploration process is further supported by Explorer by interaction mechanisms. The current set of solutions is displayed to the user as the optimization progresses, and based on this information, the user can adjust various control parameters effecting the focus of the continued search.

SAGA is an EA approach to placement introduced in [60]. Here the chosen objective is to minimize the total area of the layout while still allocating sufficient room between the blocks for the subsequent routing (see Chapter 1). SAGA is based on a unification of the EA with *Simulated Annealing* (SA). SAGA starts out as a conventional EA, but as the EA stagnates, the algorithm gradually and adaptively switches towards SA. The population size is decreased and the number of attempted mutations is increased. Mutations are carried out only with a certain probability depending on the individuals temperature, as in SA. Ultimately, the population size may decrease to one, in which case the process is pure SA.

A placement is represented by a binary tree with some additional properties. This problem-specific representation enforces the satisfaction of most constraints of the problem. The layout quality obtained by SAGA is competitive to state-of-the-art approaches at the cost of excessive, inferior runtimes. The combination of the EA with SA is found to perform significantly better than the EA alone.

Following floorplanning or placement, the area not occupied by blocks is divided into rectangular areas, called routing regions. The task of routing, i.e. implementing the connections between blocks, is divided into *global routing* and *detailed routing*. Global routing determines the global route for each net, in the form of a list of routing regions the net passes through. The objective is to minimize total layout area while not exceeding the routing capacity of each region. Global routing is performed in terms of a routing graph extracted from the given placement, and finding the global route of a single net is equivalent to solving the *Steiner Problem in a Graph* (SPG) in the routing graph. An EA for the SPG is presented in [59]. The genotype is a bitstring specifying selected Steiner vertices, and every bitstring is interpreted as a feasible solution by using a deterministic SPG heuristic, called the ***Distance Network Heuristic***, for decoding. On random graphs, the EA is superior to other heuristics for the SPG both with respect to solution quality and runtime.

A global routing algorithm is presented in [52]. It first uses the EA of [59] to generate a number of alternative, short routes for each net. I.e. the SPG algorithm is executed as many times as there are nets. Then, another EA is used to select a specific route for each net among the alternatives generated earlier, such that the total layout area is minimized. This EA uses an integer representation, where the value of the ith gene specifies the selected route for net i. The router is superior to TimberWolfMC, a state-of-the-art global router, with respect to solution quality, while not being competitive with respect to runtime.

6.3.3 Floorplanning

When determining the floorplan of an integrated circuit the objective is to find a solution which is satisfactory with respect to a number of competing criteria. At the floorplanning stage, typical criteria include layout area, aspect ratio, routability, timing and power consumption. Most often specific constraints have to be met for some criteria, while for others, a good trade-off is wanted. This situation is typical for a wide range of multi-objective problems within VLSI CAD, as discussed in general in Section 4.3. And as described in Section 4.3, the typical approach is a problem formulation based on weights and bounds such as Expression (4.1), despite the fact that this formulation might cause serious practical problems from the users point of view.

In this section an interactive floorplanner is presented which overcomes these problems by applying the preference relation introduced in Section 4.3.1. The floorplanning tool called Explorer was introduced in [57] and later improved in [58]. Explorer performs *explicit* design space exploration in the sense that

1. a set of alternative solutions rather than a single solution is generated and
2. solutions are characterized explicitly by a cost value for each criterion instead of a single, aggregated cost value, as described in Section 4.3.1.

Explorer simultaneously minimizes layout area, deviation from a target aspect ratio, routing congestion and the maximum path delay. Guided interactively by the user, Explorer searches for a set of alternative, good solutions. The notion of a "good" solution is gradually refined by the user as the optimization process progresses and knowledge of obtainable trade-offs is gained. Consequently, no a priori knowledge of obtainable values is required. From the output solution set, the user ultimately chooses a specific solution representing the preferred trade-off.

Explorer has three additional significant characteristics:

1. The maximum routing congestion is minimized, thereby improving the likelihood that the generated floorplans are routable without further modification.
2. The delay minimization is path based, while most timing-driven floorplanning and placement approaches are net-based and therefore may overconstrain the problem [71].
3. Explorer is based on evolutionary principles since the EA is particularly well suited for (interactive) design space exploration [64]. An earlier EA based approach to floorplanning is presented in [24] which, however, does not consider delay or routing congestion or explores the design space.

Problem Domain

Explorer is based on the following definition of the floorplanning problem. The input is:

1. A set of *blocks*, each of which has $k \geq 1$ alternative *implementations*. Each implementation is rectangular and either *fixed* or *flexible*. For a fixed implementation, the dimensions and exact pin locations are known. For a flexible implementation, the area is given but the aspect ratio and exact pin locations are unknown.
2. A specification of all nets and a set of paths. Capacitances of sink pins, driver resistances of source pins, internal block delays and capacity and resistance of the interconnect is also needed to calculate path delays.
3. Technology information such as the number of routing layers available on top of blocks and between blocks.

Each output solution is a specification of the following:

1. A selected implementation for each block.
2. For each selected flexible implementation i, its dimensions w_i and h_i such that $w_i h_i = A_i$ and $l_i \leq h_i / w_i \leq u_i$, where A_i is the given area of implementation i and l_i and u_i are given bounds on the aspect ratio of i, which is assumed to be continuous.
3. An absolute position of each block so that no pair of blocks are closer than a specified minimum distance. Since multi-layer designs are considered, it is assumed that a significant part of the routing is performed on top of the blocks.
4. An orientation and reflection of each block. The term *orientation* of a block refers to a possible 90 degree rotation, while *reflection* refers to the possibility of mirroring the block around a horizontal and/or a vertical axis.

IO-pins/pads are also handled by Explorer, but for brevity this aspect is not discussed here.

Evolutionary Algorithm

Explorer uses a 4-dimensional cost vector $c = (c_{area}, c_{ratio}, c_{delay}, c_{cong})$, in which each component is to be minimized:

- c_{area} is the layout area.
- c_{ratio} is the distance from a target aspect ratio.
- c_{delay} estimates the maximum path delay.
- c_{cong} is a measure of routing congestion.

The following paragraphs describe the key components of Explorer and shows how the vector-valued cost function is optimized.

Representation

The representation of a floorplan having b blocks consists of five components **1)** through **5)**:

1) A string of b integers specifying the selected implementations of all blocks. The ith integer identifies the implementation selected for the ith block.

2) A string of real values specifying aspect ratios of selected flexible implementations. The ith value specifies the aspect ratio of the ith selected flexible implementation.

3) An inverse Polish expression over the alphabet $\{0, 1, \ldots, b-1, +, *\}$ of length $2b-1$. The operands $0, 1, \ldots, b-1$ denotes block identities and $+, *$ are operators. The expression uniquely specifies a slicing-tree for the floorplan, as first introduced in [156], with $+$ and $*$ denoting a horizontal and a vertical slice, respectively.

4) A bitstring of length $2b$ representing the reflection of all blocks. The reflection of the ith block is specified by bits $2i$ and $2i+1$.

5) A string of integers specifying a critical sink for each net, to be used when routing the nets. The ith integer identifies the critical sink of the ith net.

Given a representation of the above form, the decoder computes the corresponding floorplan and its cost $c = (c_{area}, c_{ratio}, c_{delay}, c_{cong})$ in eight steps as follows:

1. The dimensions of each selected flexible block is computed from its aspect ratio and its fixed area. The dimensions of all blocks are then known.

2. From the slicing-tree specified by the Polish expression the orientation of each block is determined such that layout area is minimized. An algorithm by Stockmeyer [148] is used, guaranteeing a minimum area layout for the given slicing-structure.

3. Absolute coordinates are determined for all blocks by a top-down traversal of the slicing-tree.

4. The layout is compacted, first vertically and then horizontally. The area c_{area} is computed as the smallest rectangle enclosing all blocks and the aspect ratio cost is computed as $c_{ratio} = |r_{actual} - r_{target}|$, where r_{actual} is the actual aspect ratio of the layout and r_{target} is a user-defined target aspect ratio.

The first four steps will be illustrated by the following example:

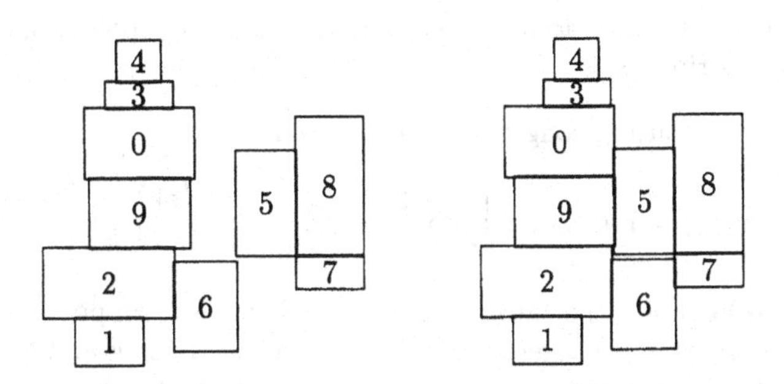

Figure 6.3 Example of 10 blocks

Example 6.4 Given 10 blocks and the Polish expression

$$1\,2+6*9\,0++\,3\,4++\,5*7\,8+*$$

the floorplan on the left in Figure 6.3 is the result of step 3. No blocks are moved when attempting vertical compaction, but subsequent horizontal compaction moves blocks 8, 7, 5, 9 and 0 towards the left, so that blocks 2, 6, 7 now determine the width of the layout. The floorplan to the right is the result of compaction (step 4).

The remaining steps are:

5. A global routing graph $G = (V, E)$ is constructed, forming a uniformly spaced lattice, covering the layout. Each pin is then assigned to the closest vertex in V. When computing this assignment, the exact pin locations are used for pins of fixed implementations, while pins of flexible implementations are assumed to be located at the center of the block.

6. The topology of each net is approximated by a Steiner tree embedded in G. Each Steiner tree is computed independently by the SERT-C algorithm ("Steiner Elmore Routing Tree with identified Critical sink") introduced in [10]. For each net, SERT-C minimizes the Elmore delay from the source to the critical sink specified by the representation.

7. The maximum path delay c_{delay} is determined by computing all path delays. For each net segment of a path, its Elmore delay is calculated in the corresponding Elmore-optimized Steiner tree and the appropriate internal block delays are added to obtain the total path delay. Since the

Steiner trees are a very accurate estimation of the net topologies, c_{delay} is an accurate estimate.

8. The maximum routing congestion is estimated as

$$c_{cong} = 100 \times \max\left[\max_{e \in E}\left\{\frac{\text{usage}(e) - \text{cap}(e)}{\text{cap}(e)}\right\}, 0\right]$$

where cap(e) denotes the capacity of edge e (depending on possible blocks at that location) and usage(e) is the number of nets using e. I.e. c_{cong} is the maximum percentage by which an edge capacity has been exceeded. The smaller c_{cong} is, the fewer nets needs to be rerouted to obtain 100% global routing completion.

Objective Function and Selection

Let Π be the set of all floorplans and $\mathbf{R}_+ = [0, \infty[$. The cost of a solution is defined by the vector-valued function $c = (c_{area}, c_{ratio}, c_{delay}, c_{cong})$ described in the previous paragraphs. As described in detail in Section 4.3.1, the user specifies neither weights nor bounds, but instead defines a goal and feasibility vector pair $(g, f) \in G$, where $G = \{(g, f) \in \mathbf{R}^n_{+\infty} \times \mathbf{R}^n_{+\infty} \mid \forall i : g_i \leq f_i\}$, $\mathbf{R}_{+\infty} = [0, \infty]$. For the ith criterion, g_i is the maximum value wanted, if obtainable, while f_i specifies a limit beyond which solutions are of no interest.

Example 6.5 The 3rd criterion minimized by Explorer is path delay. $g_3 = 5$ and $f_3 = 18$ states that a delay of 5 or less is wanted, if it can be obtained, while a delay exceeding 18 is unacceptable. A delay between 5 and 18 is acceptable, although not as good as hoped for.

The values specified by (g, f) are merely used to guide the search process and in contrast to traditional, user-specified bounds, need *not* be obtainable. Furthermore, as will be discussed in detail later, (g, f) are (re-)defined interactively at runtime. Therefore, the specification of the (g, f) vectors do not cause any of the practical problems caused by traditional weights and bounds discussed in Section 4.3.

The cost of solutions are compared using the preference relation $\prec$ defined by Definition 4.2. Using $\prec$ the solutions of a given set Φ can be ranked: $r(\phi, \Phi) = |\{\gamma \in \Phi | \gamma \prec \phi\}|$ is the *rank* of ϕ with respect to Φ, i.e. the number of solutions in Φ which are preferable to ϕ. Selection is based on this notion

```
generate (Φ):
do
    select φ1, φ2 ∈ Φ ;
    crossover (φ1, φ2, ψ) ;
    mutate (ψ) ;
    insert (Φ, ψ) ;
    if gui(update)
        adjust (g, f);
    if gui (optimize)
        hillclimber (φ, k, (g', f')) ;
    gui (display);
until gui(terminate);
output Φ0;
```

Figure 6.4 *Outline of the algorithm*

of rank. Each parent for crossover is selected at random with a probability inversely proportional to its rank.

Operators

The crossover operator as well as the mutation operator operates on each of the five components of the representation independently. While the Polish expressions are handled by highly specialized operators introduced in [24], the remaining components are handled by standard operators extensively studied in the EA literature as described in Chapter 2.

A crucial property obtained by the representation, the decoder and the genetic operators is that feasibility is preserved. I.e. only feasible representations, which can be interpreted by the decoder, are ever generated.

Algorithm

The specific EA used in Explorer is outlined in Figure 6.4. The population is denoted by $\Phi = \{\phi_0, \phi_1, \ldots, \phi_{N-1}\}$ and $\Phi_0 = \{\phi \in \Phi | r(\phi, \Phi) = 0\} \subseteq \Phi$, that is, Φ_0 is the subset of best solutions in Φ with respect to $\prec$.

The algorithm starts with an initial population consisting of randomly generated solutions. It is steady-state, so in each generation two parent individuals

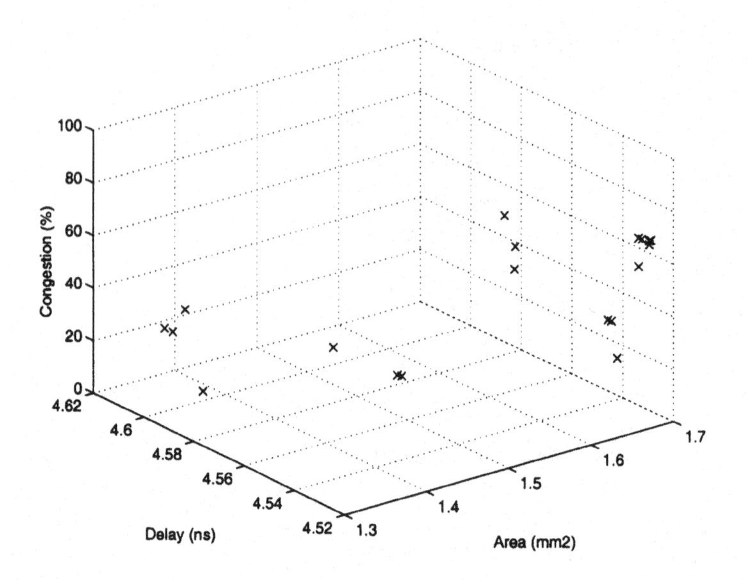

Figure 6.5 Graph showing 3 dimensions of a set of current best solutions

ϕ_1 and ϕ_2 are selected for crossover. The crossover operator generates the off-spring ψ which is then subjected to random changes by routine *mutate* and inserted into Φ by routine *insert*, replacing a poor solution. The insertion scheme ensures that a solution ψ can never replace ϕ if $\phi \prec \psi$. Hence, in the sense inferred by $\prec$ the set of best solutions Φ_0 is monotonically improving.

There are four types of interaction through the graphical user interface, or *gui*, described in the following. Explorer provides the user with continuously updated information on the current state of the optimization. The information is visualized in the form of graphs showing the cost trade-offs of the solutions obtained so far. An example graph is shown in Figure 6.5. Based on this information the user can alter the optimization process at any time by adjusting the values of (g, f) as the optimization progresses and knowledge of obtainable cost trade-offs is gained. Adjusting (g, f) redefines the notions of satisfactory and acceptable solutions. This means that the ranking of all solutions will be updated, which in turn affects the selection for crossover, i.e. the sampling of the search space. Consequently, when re-defining (g, f), the focus of the exploration process will change accordingly, allowing the user to "zoom in" on the desired region of the cost space.

Circuit	Type	Blocks	Pins	Nets	Paths
xeroxF	IC	10	698	203	86
hpF	IC	11	309	83	88
ami33F	IC	33	522	123	230
ami49F	IC	49	953	408	116
spertF	MCM	20	1,168	248	574

Table 6.5 Main characteristics of test examples.

The user can also execute a hillclimber on a specified individual ϕ. The hillclimber simply tries a sequence of k mutations on ϕ. Each mutation yielding ϕ' from ϕ is performed if and only if $\neg(\phi \prec \phi')$. The hillclimber also takes a goal and feasibility vector pair (g', f') as argument, which defines the preference relation $\prec$ to use when deciding which mutations to actually perform. This allows hillclimbing to be direction-oriented in the cost space.

The optimization process continues until the user selects termination. Before terminating, Explorer outputs the current set of distinct rank zero solutions Φ_0, i.e. the best found cost trade-offs.

Parameter settings will be described in the following section.

Experimental Results

It is inherently difficult to fairly compare the 4-dimensional optimization approach generating a solution *set* to existing 1-dimensional approaches generating a *single* solution. However, comparisons to simulated annealing and random search have been established.

Test Examples and Method

The characteristics of five of the circuits used for evaluation are given in Table 6.5. *xeroxF*, *hpF*, *ami33F* and *ami49F* are constructed from the CBL/NCSU building-block benchmarks *xerox*, *hp*, *ami33* and *ami49*, respectively, aiming at minimal alterations of the original specifications. All blocks are defined as flexible and the required timing information is added. *spertF* is a multi-chip module (MCM) designed at the International Computer Science Institute in Berkeley, California.

Explorer is implemented in *C* and executed on a DEC 5000/125 workstation. Performance is compared to that of a simulated annealing algorithm, denoted SA, and a random walk, denoted RW. Both algorithms uses the same floorplan representation and decoder as Explorer. The RW simply generates representations at random, decodes them and stores the best solutions ever found (in the $\prec$ sense). The SA generates moves using the mutation operator of Explorer and the cooling schedule is implemented following [85].

Since RW does not rely on cost comparisons, it can use the same 4-dimensional cost function as Explorer, allowing the two approaches to be directly compared. In contrast, the traditional SA algorithm relies on absolute quantification of change of cost, which therefore has to be a scalar. Using a SA cost function of the form given by Expression (4.1), it is far from clear how to fairly compare the *single* solution output by the SA algorithm to the *set* of solutions output by Explorer. Therefore, comparisons of Explorer to SA are based on optimizing one criterion only, in which case the output of Explorer reduces to a single solution.

One-Dimensional Optimization

One-dimensional optimization for area and delay was performed, for which Explorer uses the goal vectors $g = (0, \infty, \infty, \infty)$ and $g = (\infty, \infty, 0, \infty)$, respectively. Explorer is executed non-interactively.

Figure 6.6 illustrates the results. For each circuit and each of the two criteria, the three algorithms were executed 10 times each and the result indicated by a bar. The center point of each bar indicates the average result obtained in the 10 runs and the height of each bar is two times the standard deviation. For each circuit and criterion, the average result of RW is normalized to 1.00.

The SA was executed first, and the consumed average CPU time enforced on Explorer and RW as a CPU-time limit. The exact same average time consumption is thus obtained for all algorithms, at the cost of giving the SA approach an advantage. Average CPU time per run varied from 39 seconds for area optimization of *xeroxF* to about 65 minutes for delay optimization of *ami49F*. As expected, both Explorer and SA perform significantly better than RW in all cases. Overall, the performance of Explorer and SA is very similar, indicating that the efficiency of the evolutionary algorithm used by Explorer is comparable to that of SA.

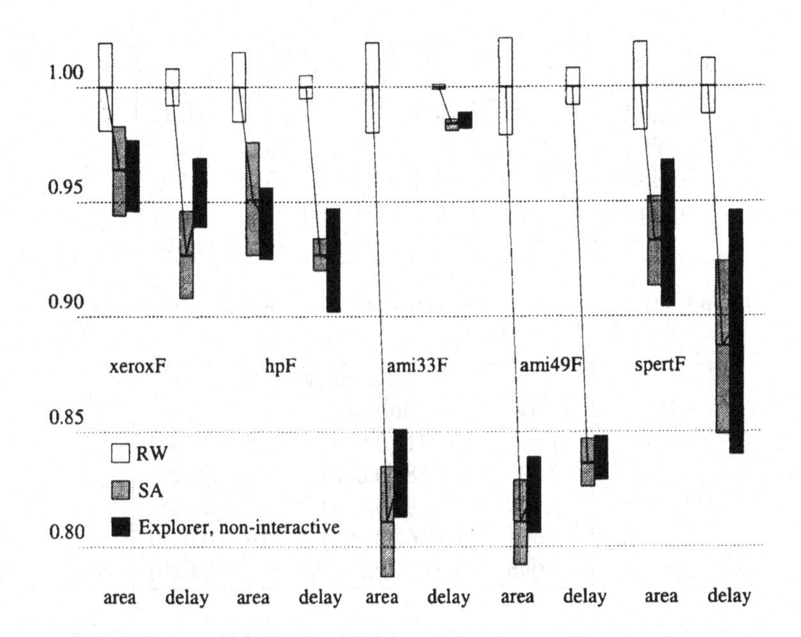

Figure 6.6 Comparison of performance for one-dimensional optimization

Four-Dimensional Optimization

Optimizing all four criteria simultaneously, interactive and non-interactive executions of Explorer are compared to RW. Explorer uses the target aspect ratio $r_{target} = 1.0$, the goal vector $g = (0, 0.2, 0, 50)$ and the feasibility vector $f = (1.5B, 0.5, \infty, 400)$, where B is the sum of the areas of all blocks of the circuit in question. For each circuit, RW is executed 10 times using a 5 CPU-hour time limit. In non-interactive mode, Explorer is also executed 10 times per circuit, but using a 1 CPU hour limit. In interactive mode, a single execution was performed for each circuit, defining the time limit as 1 hour, wall-clock time, i.e. including the time spent using the interface.

The results are shown in Tables 6.6 and 6.7. In both tables, *interact* and *non-interact* refers to the interactive and non-interactive modes of Explorer, respectively. Each entry for RW and the non-interactive mode of Explorer is the average value obtained and the value in brackets is the standard deviation. For Explorer, the output set size (Table 6.6) is limited to 40. The set quality values (Table 6.7) are obtained using the set quality measure described in Section 4.3, which accounts for the (g, f) values specified. A smaller value means a higher quality.

Circuit	Output set size				
	interact	non-interact		RW	
xeroxF	40	39.5	(1.6)	49.3	(10.1)
hpF	40	39.6	(1.0)	59.1	(14.5)
ami33F	21	34.0	(11.1)	9.7	(3.9)
ami49F	21	36.2	(4.2)	11.4	(4.5)
spertF	10	39.7	(0.7)	57.2	(17.0)

Table 6.6 Output set size of interactive and non-interactive runs versus RW

Circuit	Set quality		
	interact	non-interact	RW
xeroxF	0.572	0.741 (0.073)	0.888 (0.045)
hpF	0.605	0.638 (0.033)	0.822 (0.029)
ami33F	0.690	0.759 (0.058)	1.152 (0.048)
ami49F	0.641	0.676 (0.093)	1.197 (0.052)
spertF	0.096	1.886 (0.640)	2.178 (0.010)

Table 6.7 Set quality of interactive and non-interactive runs versus RW

The output sets obtained by Explorer in 1 hour are always significantly better than those obtained by RW in 5 hours. But more interestingly, all of the five sample executions of Explorer in interactive mode yields better results than the average non-interactive execution. Furthermore, the number of decodings performed in interactive mode averages only about 78% of that of the non-interactive mode because of the idling processor during user-interaction. Hence, using Explorer interactively significantly improves the search efficiency.

This performance gain is especially significant for the *spertF* layout. Feasible solutions were obtained interactively by executing direction-oriented hill-climbing on solutions outside but close to A_f. Only one of the sets generated non-interactively contained feasible solutions.

Conclusions

An interactive EA based floorplanner called Explorer has been presented, which minimizes area, path delay and routing congestion while attempting to meet a target aspect ratio. The key features of the approach are:

- Explicit design space exploration performed, resulting in the generation of a solution *set* representing good, alternative cost trade-offs. This is in direct contrast to traditional floorplanning approaches, which output a single solution, representing a specific compromise trade-off of the cost criteria.
- Design space exploration is enabled by using vector-valued cost rather than a traditional, aggregated scalar-valued cost function. The inherent problems of existing approaches with respect to specification of suitable weights and bounds for the scalar-valued cost function are solved by eliminating these quantities.
- An additional novel feature of Explorer is the support for user-intervention during the optimization process. The current state of the optimization is visualized for the user in the form of graphs showing the cost trade-offs of the current solutions in the cost space. Based on this information, the user can apply various interaction mechanisms, which will affect the future direction of the search. The purpose of such interaction is to allow the user to focus the design space exploration on the region of interest in the cost space.

The experimental work includes results for a real-world design. It is shown that the efficiency of the search process is comparable to that of simulated annealing and the required runtime is very reasonable from a practical point of view. Furthermore, the mechanisms provided for user-interaction are observed to improve the search efficiency significantly over non-interactive executions.

However, EA based interactive multi-objective optimization as described here is a quite new approach, and many questions remain unanswered. Further research is strongly needed, as already discussed in Section 4.3.

6.3.4 Global Routing

A well-known strategy for global routing of macro-cell layouts consists of two phases:

- In the first phase a number of alternative routes are generated for each net. The nets are treated independently one at a time, and the objective is to minimize the length of each net.

- In the second phase a specific route is selected for each net, subject to channel capacity constraints, and so that some overall criterion, typically area or total interconnect length, is minimized.

A main advantage of this routing strategy is its independence of a net ordering.

Mercury [119] and TimberWolfMC [145] are global routers for macro-cell layouts representing state-of-the-art at the time of their publication and both are based on the two-phase strategy. For nets with a small number of terminals, these routers generate up to $10 - 20$ alternative routes for each net. However, due to the time complexity of the applied algorithms, only a single route is generated for nets having more than $5 - 11$ terminals. As noted in [145] this limits the overall quality obtainable.

In this section a global router is presented which minimizes area and secondarily, total interconnect length. This router was first presented in [52]. While also being based on the two-phase strategy, it differs significantly from earlier approaches in two ways:

1. Each phase is based on an EA. The EA used in phase one provides several high-quality routes for each net independently of its number of terminals. In the second phase another EA minimizes the dual optimization criterion by appropriately selecting a specific route for each net.
2. The estimates of area and total interconnect length used throughout the optimization process are calculated very accurately. The area estimate is based on computation of channel densities and the wirelength estimate is based on exact pin locations.

Experimental results shows that the layout quality obtained by the router compares favorably to that of TimberWolfMC.

Phase One of the Router

Before the global routing process itself is initiated a rectilinear *routing graph* is extracted from the given placement. Routing is then performed in terms of this graph, i.e. computing a global route for a net is done by computing a corresponding path in the routing graph.

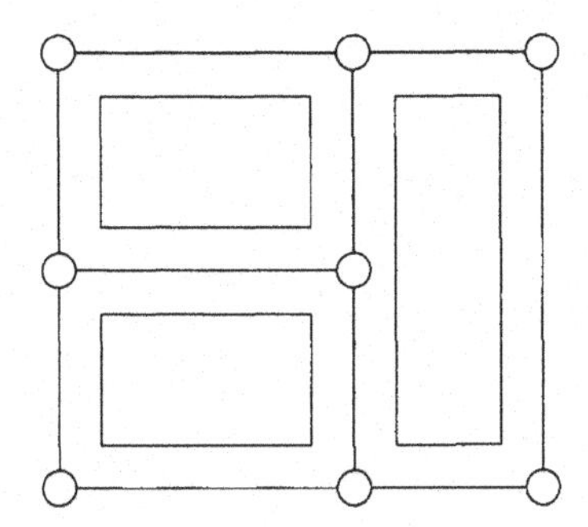

Figure 6.7 A placement and the corresponding routing graph

A quite detailed description of how to generate the routing graph for a given placement is given in [119]. Roughly speaking, each edge of the graph corresponds to a routing channel and each vertex corresponds to the intersection of two channels. An example is shown in Figure 6.7.

Before finding routes for a given net, vertices representing the terminals of the net are added to the routing graph at appropriate locations. Finding the shortest route for the net is then equivalent of finding a minimum cost subtree in the graph which spans all of the added terminal vertices, assuming that the cost of an edge is defined as its length. This problem is known as the *Steiner Problem in a Graph* (SPG). When a net has been treated, its terminal vertices are removed from the routing graph before considering the next net, thereby significantly reducing the size of the SPG instances to be solved.

For each terminal the location of the corresponding terminal vertex is determined by a perpendicular projection of the terminal onto the edge representing the appropriate routing channel, as illustrated in Figure 6.8. This is in contrast to the strategy used in e.g. [119]. Here vertices are added only at the center of routing channels and each terminal is then assigned to the closest vertex. This scheme may result in some nets having identical sets of terminal vertices, in which case some computations can be avoided. On the other hand, the scheme presented here provides a more accurate estimate of the wirelength and also allows a more accurate area estimate.

Figure 6.9 outlines phase one. A net is *trivial* if all its terminals are projected onto the same edge of the routing graph. Although several routes can still be generated for a trivial net, it will rarely be advantageous. Hence, global routing is skipped for such nets.

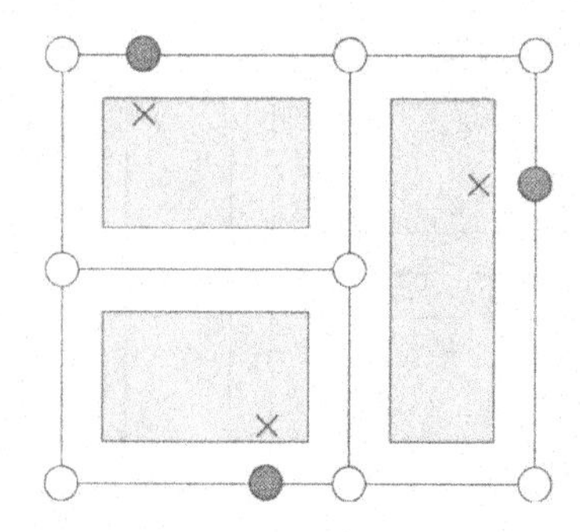

Figure 6.8 Addition of terminal vertices for a net with three terminals

generate routing graph
for (each non-trivial net)
    add vertices to graph ;
    if (2-terminal net)
        apply Lawlers algorithm ;
    else
        apply EA for SPG ;
    remove vertices from graph ;

Figure 6.9 Outline of phase one

The SPG is in general NP-complete [89]. However, if only two vertices are to be connected, SPG reduces to a shortest path problem, which is handled by an algorithm of Lawler [102]. For each two-terminal net, Lawlers algorithm is used to compute the shortest, second-shortest, third-shortest, etc. route until a maximum of R routes are found or no more routes exists. The algorithm is exact but also quite expensive, requiring time $O(R \cdot n^3)$ for one net, where n is the number of vertices in the routing graph.

An earlier algorithm presented in [50] may at first seem more attractive. It generates the R shortest routes from a designated vertex to each of the other vertices in time $O(R \cdot n \cdot log(n))$. However, loops are allowed in a path, as opposed to Lawlers algorithm, and if two paths do not visit the same vertices in the same order they are considered distinct. One could then simply generate routes until R loopless routes were obtained, which were also distinct in the sense that their sets of edges are distinct. However, experiments have shown that this strategy is not feasible in practice due to the number of routes then required.

For each net with three or more terminals, at most R distinct routes are generated using an EA for the SPG. This algorithm was first introduced in [59] and was later improved significantly [53, 54]. For a detailed description of the algorithm the reader is referred to these papers. There are two main advantages of using this algorithm in the context of the global router. Firstly, it generates high-quality solutions. In [53] the EA is tested on graphs with up to 2,500 vertices and is found to be within 1% from the global optimum solution in more than 90% of all runs. The routing graph of a macro-cell placement with C cells will have less than $3 \cdot C$ vertices. It is therefore most likely that the EA will find the shortest existing route for every net in any reasonably sized macro-cell layout. The second advantage of the EA is that it provides a number of distinct solutions in a single run. The problem of Mercury and TimberWolfMC that only one route is generated for nets with many terminals is thus eliminated.

For nets with few terminals, say 6-7 or less, exhaustive search for the shortest route will often be feasible. Using an algorithm by Sullivan [149] optimum can be found by exhausting a search space consisting of

$$\sum_{i=0}^{k} \binom{n}{i}$$

points, where $k = \min(t-2, n)$ and t is the number of terminals of the net. However, experiments have revealed that Sullivans algorithm often considers fewer distinct solutions and is slower than the EA. Therefore, the EA is used for every net with more than two terminals[2].

Evolutionary Algorithm

This section describes the EA used in the second phase of the global router. In addition to presenting the EA components, the area estimate is especially important to this specific algorithm and is discussed in detail.

[2]To obtain as many distinct solutions as possible, the EA does not use the reduction of the search space described in [53, 54].

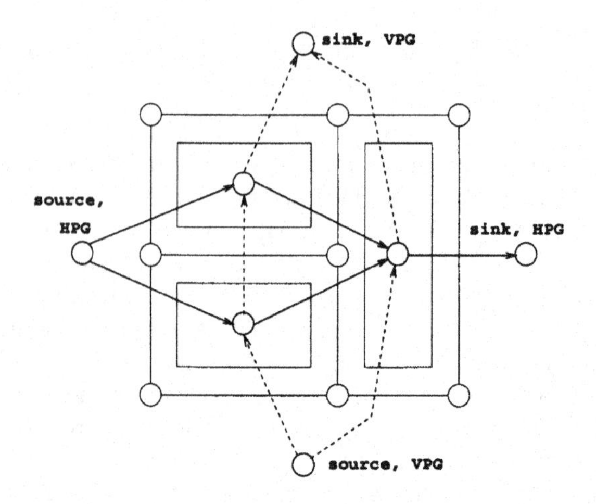

Figure 6.10 Polar graphs for area estimation

Representation

A global routing solution is represented by specifying for each net which of its possible routes is used. Assume a fixed numbering $0, 1, \ldots, N-1$ of the nets, let $\pi : \{0, 1, \ldots, N-1\} \mapsto \{0, 1, \ldots, N-1\}$ be a bijection and denote by $r_k \leq R$ the number of routes generated in phase one for the kth net. An individual is then a set of N tuples:

$$\{(\pi(0), q_{\pi(0)}), (\pi(1), q_{\pi(1)}), \ldots, (\pi(N-1), q_{\pi(N-1)})\},$$

where $1 \leq q_k \leq r_k$ for all $k = 0, 1, \ldots, N-1$. For example, the tuple (3,7) specifies that the 3rd net uses its 7th route. The mapping π defines an ordering of the nets which allows an inversion operator to be applied. Note that the routing solution specified by an individual is independent of π.

Objective Function and Selection

As in [119, 154] the area estimation is based on the formation of polar graphs as illustrated in Figure 6.10. For a given placement and routing graph, two polar graphs are constructed, a horizontal (HPG) and a vertical (VPG). First consider HPG. The vertices of HPG consists of a vertex for each cell plus two additional vertices, a source and a sink. Each edge in HPG corresponds to a vertical edge in the routing graph and is directed from the source towards the sink.

Assume that each edge (v, w) has a cost which corresponds to the spacing needed between cells v and w to perform the routing. Furthermore, assign to each path from source to sink a fixed cost which is the sum of the horizontal length of all cells visited on the path. The total cost of the longest path in HPG then estimates the horizontal length of the layout. By constructing VPG in a similar way, the area is estimated as the product of the longest path in HPG times the longest path in VPG.

In [119] the cost of an edge in the polar graphs is a rather simple function of the number of nets present in the corresponding routing channel. However, if m nets are present in a channel, the channel density can be any number between 0 and m, assuming that two metal layers are available for routing and that each layer is used exclusively for routing in a specific direction. Therefore, to obtain a more accurate area estimate, exact channel densities are computed for each edge in the routing graph. This is possible since the routing in phase one was performed using accurate positions for the terminals of each net. The cost of an edge in the polar graphs is then proportional to the density of the corresponding channel.

Several factors affect the accuracy of the area estimate. The two most important have to do with the subsequent compaction/spacing of the layout:

1. If the compactor alters the placement to the extent where the topology of the routing graph is changed, the polar graphs are also changed. Hence, the quality of the area estimate decreases significantly or may even become meaningless. In other words, a good initial placement is required so that the compactor will only perform minor adjustments of the cell positions. This situation reflects the well-known strong mutual dependency of the placement and global routing tasks.
2. It is implicitly assumed that the compactor generates a layout in which no routing channel on a longest path of a polar graph is wider than needed. Otherwise, the area will be underestimated.

The practical consequences of these assumptions are addressed later in this section.

For each individual in a population, its estimated area is computed as described above and its estimated total wirelength is computed as the sum of the length of the selected routes. Fitness is then computed by sorting the population $P = \{p_0, p_1, \ldots, p_{M-1}\}$ lexicographically using area as most significant criterion and wirelength as a secondary criterion. Assume that P is sorted in decreasing order with respect to this ordering. The fitness F of p_i is then computed as $F(p_i) = 2i/(M-1)$ for $i = 0, 1, \ldots, M-1$. This ranking assures constant variance of fitness throughout the optimization process. Ranking aims at controlling the speed of convergence, including avoiding premature convergence.

Operators

Given two parent individuals α and β, the crossover operator generates two offsprings ϕ and ψ. The parent individuals are not altered by the operator. In the following a (second) subscript specifies which individual the marked property is a part of. Crossover consists of two steps:

1. One of the parents, say β, is chosen at random, and a copy γ of β is made. γ is then reordered so that it becomes homologous to α, that is, $\pi_\gamma = \pi_\alpha$.
2. The offsprings are given the same ordering as their parents: $\pi_\phi = \pi_\psi = \pi_\alpha$. Standard 1-point crossover is then performed: A crossover point x is selected at random in $\{0, 1, \ldots, N-2\}$. The selected routes of ϕ is then defined by $q_{\pi(k),\phi} = q_{\pi(k),\alpha}$ if $k \leq x$ and $q_{\pi(k),\phi} = q_{\pi(k),\gamma}$ otherwise, where $\pi = \pi_\alpha$. Similarly, the selected routes of ψ is defined by $q_{\pi(k),\psi} = q_{\pi(k),\gamma}$ if $k \leq x$ and $q_{\pi(k),\psi} = q_{\pi(k),\alpha}$ otherwise.

The mutation operator performs pointwise mutation: It goes through the N tuples of the given individual and randomly selects another route for the kth net with probability $p_{mut}(r_k - 1)$, where p_{mut} is the userdefined mutation probability.

As mentioned earlier a given global routing solution can be represented by several equivalent individuals because of the independence of the ordering π. However, the fitness of offspring produced by crossover depends on the specific orderings of the given parent individuals. The purpose of inversion is to optimize the performance of the crossover operator. With a given probability p_{inv}, the inversion operator alters the ordering π of a given individual. To obtain a uniform probability of movement of all tuples, the set of tuples is considered to form a ring. A part of the ring is then selected at random and reversed. More specifically, two points $x, y \in \{0, 1, \ldots, N-1\}, x \neq y$, are selected at random.

```
generate (P_C);
evaluate (P_C);
s = bestOf(P_C) ;
repeat until stop_criterion()
    P_N = ∅;
    repeat M/2 times
        select p1 ∈ P_C, p2 ∈ P_C ;
        crossover (p1, p2, c1, c2) ;
        P_N = P_N ∪ {c1, c2} ;
    evaluate (P_C ∪ P_N);
    P_C = reduce(P_C ∪ P_N);
    ∀ p ∈ P_C : possibly mutate(p) ;
    ∀ p ∈ P_C : possibly invert(p) ;
    evaluate (P_C) ;
    s = bestOf(P_C ∪ {s});
optimize (s);
output s ;
```

Figure 6.11 Outline of phase two

The operator then defines the new ordering π' as[3]

$$\pi'((x+i) \bmod N) = \pi((y-i) \bmod N) \quad \text{if} \quad 0 \leq i \leq (y-x) \bmod N$$

and

$$\pi'((x+i) \bmod N) = \pi((x+i) \bmod N) \quad \text{otherwise} \quad i = 0, 1, \ldots, N-1.$$

Algorithm

Figure 6.11 outlines the phase two algorithm. Initially, the current population P_C of size $M = |P_C|$ consists of $M-1$ randomly generated individuals and a single individual consisting of the shortest route found for each net. Seeding the population with this relatively good solution does not lead to better final results, but merely speeds up the search process. Routine *evaluate* computes the fitness of each of the given individuals, while *bestOf* finds the individual

[3]The definition of π' relies on the mathematical definition of modulo, in which the remainder is always non-negative.

with the highest fitness. The best individual s ever seen is preserved. Routine ***stop_criterion*** terminates the simulation when no improvement has been observed for S consecutive generations. Each generation is initiated by the formation of a set of offspring P_N of size M. The two mates p_1 and p_2 are selected independently of each other, and each mate is selected with a probability proportional to its fitness. Routine ***reduce*** returns the M fittest of the given individuals, thereby keeping the population size constant. Routine ***optimize*** (s) performs simple hillclimbing by executing a sequence of mutations on s, each of which improves the fitness of s. The output of the algorithm is then the solution s.

Parameter Settings

The same set of control parameters are used for all program executions described in the following section. That is, no problem specific tuning is performed. For each net, at most $R = 30$ alternative routes are generated. The parameters of the EA used in phase one are as given in [59]. The phase two EA is executed with population size $M = 40$, stop criterion $S = 100$, mutation probability $p_{mut} = 2.5 \times 10^{-4}$ and inversion probability $p_{inv} = 0.1$.

Experimental Results

The router has been implemented in the *C* programming language, and all experiments are performed on a Sun Sparc IPX workstation. The router is interfaced with the macro-cell layout system Mosaico, which is a part of the Octtools CAD framework developed at University of California, Berkeley. This integration allows for comparison of the routers performance to that of TimberWolfMC [145], a widely known global router also interfaced to Mosaico.

Test Examples

Three of the MCNC macro-cell benchmarks, *xerox*, *ami33* and *ami49*, were used for the experiments. However, due to a purely technical problem, it became necessary to remove all pads from these examples before using them[4]. The modified benchmarks are referenced using a '-M' suffix.

Table 6.8 lists the main characteristics of the test examples. The number of nets and the number of terminals listed are totals, i.e. they include the few trivial nets. *xerox-M*, *ami33-M* and *ami49-M* are placed by Puppy, a placement

[4] In the available version of Octtools (5.2) the channel definition program Atlas could not handle the pad placement generated by Padplace.

Problem	#cells	#nets	#terms.
xerox-M	10	203	696
ami33-M	33	85	442
ami33-2-M	33	85	442
ami49-M	49	390	913
ami49-2-M	49	390	913

Table 6.8 Problem characteristics

tool based on simulated annealing, also included in Octtools. *ami33-2-M* and *ami49-2-M* are other placements of *ami33-M* and *ami49-M*, respectively. The generation of these placements are described below.

Method

Two factors makes it difficult to device a sequence of experiments providing an absolute fair performance comparison of the two global routers. Firstly, global routing is just one of a sequence of heavily interacting steps needed to generate a complete layout. Hence, when considering a specific result, it may be influenced by a pattern of interactions with other tools, which accidentally favours one of the routers. Secondly, the optimization strategies used by the two routers are not identical. As described earlier, the EA based router explicitly attempts to minimize area and secondarily wirelength. While TimberWolfMC also generates the shortest possible routes in phase one, area is not an explicit component of the optimization criterion used in the second phase. Instead, TimberWolfMC selects the shortest possible routes subject to channel capacity constraints.

The chosen strategy for experiments are as follows: For each of the placed examples listed in Table 6.8, Mosaico was executed to generate a complete layout, using either TimberWolfMC or the EA based router for the global routing task. Hence, all other steps of the layout process are performed by the same tools.

Mosaico was executed five times for each example using the EA based global router in order to capture the variations caused by the stochastic nature of the applied algorithms. There was no variation of results when applying Timber-WolfMC.

Problem	Solution	A_{tot}	A_{route}	WL
xerox-M	best	−1.9	−4.7	+0.0
	avg	−1.4	−3.5	+0.8
ami33-M	best	−3.2	−5.1	−3.2
	avg	+1.6	+2.5	−0.2
ami33-2-M	best	−3.0	−4.7	−1.5
	avg	−1.1	−1.7	−0.2
ami49-M	best	−1.9	−3.3	−1.5
	avg	−0.5	−0.8	+0.3
ami49-2-M	best	−4.2	−7.3	−4.0
	avg	−3.7	−6.3	−2.9

Table 6.9 Relative improvements obtained by the EA based router

Layout Quality

Table 6.9 summarizes the impact on the completed layouts of using the EA based router instead of TimberWolfMC. A_{tot} denotes total area, A_{route} denotes routing area, i.e. the part of the total area not occupied by cells and WL denotes total wirelength. Each entry is computed as 100(EA-result/TW-result − 1). Hence, a negative value indicates a reduction in percent obtained by the EA based router, while a positive value indicates a percentage overhead as compared to TimberWolfMC. Despite the inherent problems of this kind of comparison discussed above, it is clear that in general the EA based global router obtains the best layout quality for the problem instances considered.

Inspection of the generated layouts reveals interesting information regarding the two major assumptions underlying the area estimation. The placement of *xerox-M* is adjusted only slightly during compaction, and the routing graph topology is unaltered. For this example, the EA based router obtains an average reduction of 3.5% of the routing area which comes at the price of a 0.8% increase in total wirelength. However, for *ami33-M*, the EA based router on average obtains larger layouts than TimberWolfMC. In this case the placement, and hence the routing graph topology, is significantly modified by the compactor. As a consequence, the function minimized by the EA based router in its second phase correlates very poorly to the actual layout generated, which inevitably leads to a poor result. To counteract this phenomenon, a new placement *ami33-2-M* was produced by ripping up all routing in the completed layout of *ami33-M* generated using TimberWolfMC. Since the placement thus obtained is the result of compaction and completion of all routing, it will probably only be subjected to minor adjustments when used itself as input to Mosaico. Experiments confirmed this assumption. The topology of the routing graph of *ami33-2-M* is unaltered throughout the process and the performance of the EA based router is now superior to that of TimberWolfMC. Very similar results are observed for *ami49-M*: The routing graph topology is significantly altered during the layout process. The placement of *ami49-2-M* is obtained the same way as *ami33-2-M*, and the performance of the EA based router improves significantly on this example.

The significant routing graph alterations for some problems are a consequence of rather poor initial placements. It is not clear how better placements would affect the relative performance of the two routers. As placement quality increases, the relative effect of eliminating a wire from the longest path in a polar graph increases, indicating a potential advantage for the EA based router. On the other hand, a good placement contains less routing, suggesting that the performance gap would be narrowed.

For the test examples considered here, most routing channels on the longest paths are compacted to their minimum widths by the compactor (see the second assumption discussed earlier). However, in most cases at least one channel on the longest paths are still wider than necessary. Hence, the area estimation performed tends to underestimate the final area. However, this second assumption appears to be fairly reasonable.

Runtime

On average the router requires about 22, 12 and 130 minutes to route examples based on *xerox*, *ami33* and *ami49*, respectively. TimberWolfMC spends about 30 seconds for examples based on *xerox* and *ami33*, and about 5 minutes for *ami49* based examples. Hence, the EA based router is clearly inferior to TimberWolfMC with respect to runtime. The total layout generation process performed by Mosaico (i.e. excluding placement) requires about 15 minutes for examples based on *xerox* and *ami33*, and about an hour for *ami49* based examples, when TimberWolfMC is used. Hence, the use of the EA based router increases the layout generation time by a factor of two or three.

However, the runtime of the current implementation can be improved significantly in a number of ways. The vast majority of the runtime is spend computing channel densities. When estimating the area of a solution, all densities are recomputed whether the routing in a channel is actually changed or not. Keeping track of the need to recompute channel densities and updating channel densities dynamically would hence reduce runtime significantly. Another approach is to implement a parallel version of the router. Due to the inherent parallelism of any EA, a high speedup can be expected on any MIMD architecture.

Conclusions

In this section a novel approach to global routing of macro-cell layouts based on EAs has been presented. The key features and conclusions are as follows:

- The approach contains two main phases, each of which are based on an EA. In the first phase, an EA for the Steiner tree problem in a graph is used to generate a set of alternative routes for each net. Then, in the second phase, another EA selects a specific route for each net, considering both the estimated layout area and the total wirelength.

- This application illustrates how the complete solution set maintained by an EA can be of explicit use in certain contexts. While only the best solution obtained by an EA is typically output and used, the phase two EA presented here explicitly utilizes not only the best solution found for each net by the phase one algorithm, but also the second best, third best, etc. The best overall layout area will often be obtained by *not* using the very best solutions from phase one (the absolute shortest route for each net), but instead using longer routes for some nets. In other words, the information represented by the solution set of the phase one algorithm is utilized fully in this context.

- The performance of the router is compared to that of TimberWolfMC on MCNC benchmarks. The experimental setup illustrates how closely related the placement and global routing tasks are. They are mutually dependent in various ways, which makes it difficult to completely isolate for example the global routing problem and solving and evaluating it in isolation. Such mutual dependencies are typical for a wide range of VLSI CAD problems.

- Experimental results shows that the quality of completed layouts improves when using the EA based router instead of TimberWolfMC, assuming that the quality of the given placement is sufficiently high. The router is inferior to TimberWolfMC with respect to runtime, but major improvements are possible.

Since the work presented here is a first approach to global routing based on EAs, future improvements of the layout quality obtainable are also likely.

6.3.5 Detailed Routing

If we consider the "natural" flow in the CDP the next problem occuring during the physical design phase of ICs is the *Detailed Routing Problem* (DRP).

In the area of detailed routing a main task is to connect pins of signals within a routing region, given the global routing. Each net is implemented by a number of connected wire segments, assigned to metal layers of the chip. (If a rectangular routing region has pins only along two parallel sides, it is called a *channel.*) The net implementation has to obey a large set of constraints defined by the specific technology, like the underlying routing model and the number of available layers. Furthermore, the designer has to consider quality factors for the resulting layout. The routing area should be "small" to minimize the chip area. Also delay times are given in the specification which must also be satisfied. There are further aspects which must be considered, like testability properties, propagation delay, and undesirable tensions (*crosstalk*) on a manufactured circuit. Additionally, the fabrication costs are an aspect that has to be considered during the design process of an IC. In the last years various approaches based on EAs for detailed routing have been presented (see e.g. [146, 109, 127, 106, 104, 105]). But most EA approaches fail for large examples due to their runtimes and memory requirements, e.g.: An EA for channel routing is presented in [106]. The main criterion optimized is the channel area, and a secondary criterion is the total netlength. The genotype is essentially a lattice corresponding to coordinate points of the layout, and does not seem to support the genetic operators that well. Crossover is performed in terms of wire segments. A randomly positioned line perpendicular to the pin sides divides the channel into two sections and plays the role of the crosspoint. Wire segments exclusively on, say, the left side of the cross-line are inherited from the first parent, and segments exclusively to the right of the cross-line are inherited from the second parent. Segments intersecting the cross-line are substituted in the offspring by new randomly generated segments. Similarly, mutation consists of replacing selected segments by random ones. On benchmark examples, the router produces results equal to or better than the previously best published results, while not being runtime competitive.

In the following an EA for the DRP described in [78, 76] is presented. Even though the underlying ideas of the approach are similar to [106] the approach avoids several drawbacks:

- The algorithm uses an encoding that also allows (temporarily) invalid solutions. By this, a lot of time is saved since the time consuming repair runs do not have to be carried out each time. The invalid solutions are "sorted out" by the fitness function that is chosen accordingly.
- Often it is useful to combine "pure" EAs with problem specific knowledge. Here a rip-up and reroute technique (that can be seen as a greedy algorithm) is used to create the initial population and the genetic operators are established on problem specific methods.
- The algorithm can handle multi-layer channels. Furthermore, several optimization goals, like net length, number of vias, crosstalk, are considered in parallel. The algorithm can be applied not only to channels, but also to switchboxes, i.e. routing regions with pins located on all sides of the borders.

Experiments are reported that demonstrate the efficiency of the EA approaches.

Detailed Routing Problem

In the following the *Detailed Routing Problem* (DRP) under a set of constraints is briefly described (see also [103]). Furthermore, basic definitions that are used below are introduced. Then some basic properties and algorithms are discussed.

- A *switchbox* is a rectangular routing region (routing area) which is located between circuit blocks that are to be connected. (Notice, that due to the obstacles that are allowed also non rectangular structures can be considered.)
- The routing has to be performed on horizontal and vertical wires that are located in the switchbox. A horizontal row is called a *track* and a vertical wire is named *column*.
- The number of tracks defines the *width* of the switchbox.

- Furthermore, *pins* are allowed to be located on all side of the switchbox and represent the connection points on circuit blocks.
- A *net* is the union of pins that have to be connected and wire segments (tracks and columns) that are used to connect these pins.

The same definitions hold for a *channel* with the restriction that pins are only located at the upper and lower side and that the number of tracks in the channel is not fixed, i.e. in this case the *Channel Routing Problem* (CRP) is to find a solution within a minimum number of tracks.

Several restrictions are considered in the following:

- The underlying detailed routing model is the *Manhattan Model* [103]. Therefore different nets have to be disjoint, i.e. it is not allowed that two nets occupy the same wire segment.
- Several different interconnection layers are available for routing, but it is assumed that all pins are located on the lowest layer (*surface connection*).
- If a net is realized on different layers the connection between the two layers is called a *via*.
- Crosspoints of two different nets in the same layer are named *violations*.

Remark 6.1 A feasible routing solution has no violations.

Furthermore, we need the definition of the *Manhattan Distance* [103]. It is defined as the minimum net length between two pins.

In the following five optimization goals are considered to model the constraints given by the DRP and the technology. They are specified by (see also [87]):

Complete Routing Solution: The primary goal of the EA based router is to find a feasible routing solution, where each net is fully connected and no violations are present.

Width: The number of tracks which are needed to realize a routing solution has to be minimized to obtain a minimum routing area.

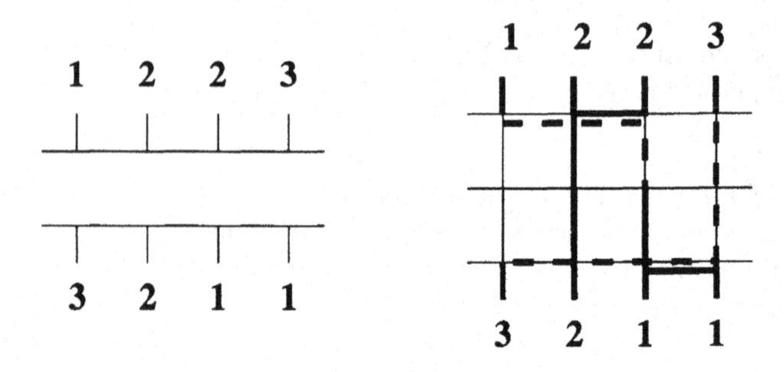

Figure 6.12 Instance of two-layer CRP

Net length: The net length of the routing solution is defined by the sum over the length of all nets. The length of a net largely influences the propagation delay of the realized circuit.

Number of vias: Each via directly requires costs in fabrication. Furthermore, it forces a higher propagation delay and increases the defect probability in the resulting circuit.

Crosstalk: The crosstalk effect results from wire segments which are located on the same track segment in adjacent layers. Thus, also the number of wire segments which are located in parallel on adjacent layers is minimized.

The optimization goals are considered simultaneously, i.e. multi-objective optimization is performed (see Chapter 4). The importance of each of the quality factors in the optimization process is controlled by a weighted sum which is described below.

Example 6.6 In Figure 6.12 an instance of a two-layer routing channel (a special case of the DRP) is illustrated. The left picture shows the channel with pins that are to be connected. Pins belong to the same net if they are denoted by same integer numbers. The right picture shows a routing solution with two layers and channel width $w = 3$. Solid lines represent the connections in the lower layer and dashed lines represent the connections in the upper layer.

For the routing model considered several layers are available and crosstalk should be avoided, but in general it is allowed.

If the number of tracks in the routing area is not fixed the initialization algorithm of the EA starts with computing a lower bound for the track number.

Let I be an instance of the DRP and let $\{1,\ldots,m\}$ be the set of columns. Following [103] a lower bound for the number of tracks can be calculated by the following equation:

$$\widetilde{W}(I) = \max_i |\mathcal{N}_i| / \#layers, \tag{6.3}$$

where $i \in \{1,\ldots,m\}$ and $\mathcal{N}_i$ is the set of nets that have at least one pin in a column $\leq i$ and another in a column $\geq i$. This lower bound is used to determine the number of tracks that are at least needed for finding a valid solution. By this not too much runtime is wasted during initialization of the EA.

Finally, a rip-up and reroute technique is introduced: The genetic operators of the EA make use of domain specific knowledge and methods that are well-known in routing approaches. One of them is to *rip-up and reroute* nets in a routing region.

Rip-up and Reroute: The heuristic starts from a completely wired solution, where each net is fully connected. The solution must not necessarily be feasible, i.e. violations may occur. Then one net η, $\eta \in \mathcal{N}$, is deleted ("rip-up") in the solution. The unconnected pins are connected using a heuristic ("reroute") that is introduced in the next section.

Evolutionary Algorithm

In this section the EA is described that is applied to the DRP. The structure of the EA and the functionality of the genetic operators are discussed in detail. Finally, the parameter settings are given which are the same for all problem instances considered later.

Representation

A multi-valued encoding is used to represent a solution of an instance of the DRP. The elements are given as finite strings which have no fixed length.

The representation scheme works as follows: For each element the information is stored which wire segments in the routing channel are occupied by net η, $\eta \in \mathcal{N}$, where $\mathcal{N}$ denotes the set of nets. The length of the strings is then limited by

$$C \cdot \sum_{\eta \in \mathcal{N}} length(\eta),$$

where C denotes a constant factor for handling the representation. Notice, that the length of the strings is limited by the sum over all nets inside the channel and not by the size of the channel. Therefore, also large channels can be handled.

A population is a finite set of elements as defined above. To the elements various genetic operators are applied that will be introduced below.

Objective Function and Selection

The *objective function* calculates the fitness of each element in the population. Let x be an element of the population. First, for each net η, $\eta \in \mathcal{N}$, which occurs in x a fitness value is determined by the function $F(\eta)$ given by:

$$F(\eta) = \frac{a \cdot length(\eta) + b \cdot vias(\eta) + c \cdot violations(\eta) + d \cdot crosstalk(\eta)}{\sum_{pins \in \eta} manhattan\_distance} \quad (6.4)$$

$length(\eta)$ denotes the net length of net η and $vias(\eta)$ ($violations(\eta)$) counts the number of vias (violations) of net η. $crosstalk(\eta)$ counts the number of parallel wire segments on adjacent layers. The weights of the sum are $a, b, c, d \in \mathbf{R}$ and *manhattan_distance* is the sum of the manhattan distance of all pairwise pins of net η. (For a more detailed discussion on the advantages and disadvantages of using a weighted sum as a fitness function see Chapter 4.)

Notice that infeasible solutions may occur in the present population. Each violation of an element is counted by the penalty term $c \cdot violations(\eta)$. (More details of the choice of the factors a, b, c and d are given below.) The fitness value $F(\eta)$ for each net η, $\eta \in \mathcal{N}$, is then divided by the minimized manhattan distance of the considered net; by this the fitness is normalized.

Using the formula described above the fitness of an individual x is then calculated by

$$fitness(x) = \sum_{\eta \in \mathcal{N}} F(\eta). \quad (6.5)$$

Thus, by evaluating the objective function we follow the optimization goals

1. to find a feasible solution, i.e. one without violations, and
2. to minimize the costs for the realization of the routing solution, i.e. the minimization of the net length, the number of vias, and the crosstalk effect.

During the EA run individuals are selected for creating offsprings by applying genetic operators. Again, the selection is performed by ***roulette-wheel selection*** and ***steady-state reproduction***.

Initialization

First, the initialization algorithm INIT of the EA is described, that constructs the initial population of size $|\mathcal{P}|$. INIT consists of a heuristic that chooses the routes in the routing region. The heuristic tries to choose the shortest connection between two pins. If this is impossible more and more detours are admitted. This heuristic is called *Random Router* and works as follows:

Random Router (RAND): Let $\{pin_1, \ldots, pin_m\}$ be a set of pins that have to be connected by the same net η. Begin with pin pin_1 and successively connect it to the other pins pin_i ($i \in \{2, \ldots, m\}$) by choosing one of (at most) six possible directions (these are: *left, right, forward, backward, up* and *down*) in the channel.

RAND works in several modes. Each mode in RAND depends on the choice of the probabilities for each direction. For the lowest mode M_0 in RAND the probability for the choice of the *shortest* direction in the channel is set to 1 and the remaining directions are set to 0. (Notice, that probability 1 for the shortest direction which goes straight ahead to the next pin and probability 0 for each other direction implies an algorithm to obtain a *minimized manhattan distance.*) The different modes M_i, $1 \leq i \leq max$, are adapted as follows: If i is increasing other directions than the shortest one become more and more likely. The number of different modes for RAND is constant (see discussion of parameters below).

By using RAND the initialization algorithm INIT works as follows:

Initialization (INIT): Take the nets of the routing region in random ordering. Construct the first net by RAND in mode M_0. Then increment the

```
initialization_algorithm (region):
  for (each net)
    do
        random_router (mode) ;
        if (violation)
            increment (mode) ;
        if (mode = max)
            random_router_violation (max) ;
    while (net not routed) ;
  return ;
```

Figure 6.13 Sketch of the initialization algorithm

mode of the algorithm RAND to M_1 and perform RAND in this new mode until a violation appears. In this case increase i, $1 \leq i \leq max$, and apply RAND again until M_{max} is obtained. If a legal routing solution, i.e. one without violations, is not obtained perform the routing with RAND in mode M_{max} and allow the occurrence of violations.

A sketch of the algorithm is given in Figure 6.13.

To construct an initial population the algorithm INIT is performed $|\mathcal{P}|$ times with different random orderings of the nets.

Genetic Operators

In the following the genetic operators are introduced. Each operator works on the encoded representation of the routing solutions and creates new elements. During the EA run violations are allowed, i.e. the operators need not necessarily create feasible solutions.

In the application two recombination operators for the creation of new individuals are used. Both methods create offsprings from two parents. The parent elements are selected as described above.

The first crossover operator needs an algorithm that calculates a cut position for crossing the elements:

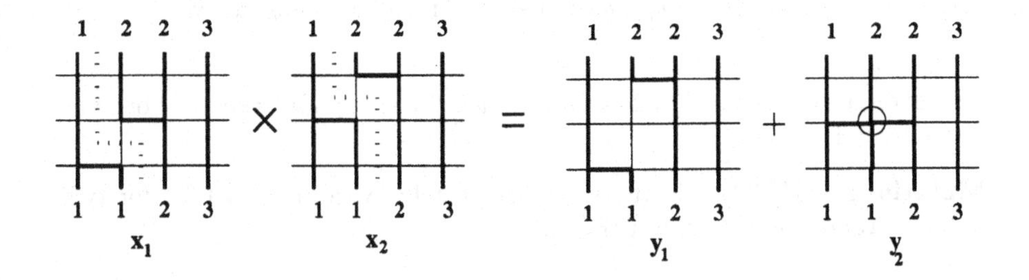

Figure 6.14 Illustration of crossover

CUT: Choose a net η, $\eta \in \mathcal{N}$, at random and set the cut position of the parent elements behind net η. This corresponds to an integer cut position d in the string representation. (Notice, that the crossover is applied to strings and therefore an integer value must be calculated as cut position.)

Then the two offsprings are constructed as follows:

Crossover: Create two new elements y_1 and y_2 from parents x_1 and x_2. Split each element x_1 and x_2 in two parts at a cut position d chosen by function CUT. The first part of child y_1 (y_2) is taken from x_1 (x_2) and the second part of y_1 (y_2) is taken from x_2 (x_1).

Example 6.7 Figure 6.14 shows an example of the crossover operator where only one layer is illustrated. The cut position is chosen behind net 1. Offspring y_1 has a shorter netlength than the parents, while offspring y_2 is an infeasible element.

Furthermore, a recombination operator is used that merges two parents to get one offspring:

Merge Crossover: Create one new element y from parents x_1 and x_2. y takes each net η, $\eta \in \mathcal{N}$, one after another from the parent, whose fitness $F(\eta)$ is of higher quality.

Two mutation operators are applied that also work on the string representation. They are based on the ***rip-up and reroute*** technique described above.

Mutation1 (MUT1): Create a new element by ***rip-up and reroute*** of one net randomly chosen.

Mutation2 (MUT2): Create a new element by performing MUT1 for two nets randomly chosen in parallel.

Algorithm

The EA computes feasible routing solutions for a given routing region and a fixed number of layers. Using the genetic operators it works as follows:

- The algorithm starts with a fixed number of tracks. It is calculated by Equation (6.3).
- The initial population is generated by applying INIT.
- Each element in the initial population is evaluated using the objective function given by Equation (6.5).
- In each iteration two parents are selected and one of the recombination operators is applied. Then one of the two mutation operators is selected with corresponding probability and applied to a randomly chosen element.
- The newly created offsprings are evaluated by the objective function.
- If no improvement is obtained for 5000 generations and a solution without violations is computed the algorithm stops. Otherwise, for each element one track is inserted into the channel. The elements are enlarged where the channel density is the highest (analogously to [106]). Then the EA is restarted. By this strategy it is guaranteed that a valid solution is found.
- Finally, the elements are optimized by a post processing. A local netlength optimization is then performed for the best element. For this detours are eliminated by one pass over the routing region. This method simplifies structures analogously to Figure 6.15.

A sketch of the algorithm is given in Figure 6.16.

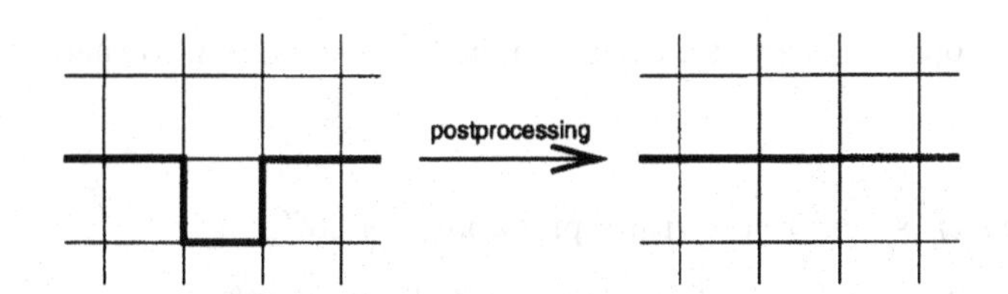

Figure 6.15 Elimination of unnecessary detours

```
evolutionary_algorithm (region, # layers):
    generate_initial_population ;
    calculate_fitness ;
    do
        apply_operators_with_corresponding_probabilities ;
        calculate_fitness ;
        update_population ;
    while (not terminal case) ;
    post_processing ;
    return best_element ;
```

Figure 6.16 Sketch of the EA

Parameter Settings

The number of modes for RAND in INIT is set to 6. (Thereby a compromise between runtime of the initialization algorithm and quality of the initial population is obtained.) The size of the population is set to 10, since for larger population sizes runtimes are no longer in an acceptable range.

The genetic operators are iteratively applied corresponding to their probabilities:

1. *Crossover* is performed with a probability of 60%.
2. *Merge Crossover* is applied with a probability of 10%.
3. Each mutation operator is applied with a probability of 20%.

For the objective function the factors a, b, c and d are chosen $a = 1$, $b = 10$, $c = 1000$ and $d = 10$. The choices of the weights are based on experiments. The primary optimization goal is to obtain feasible solutions during the run of the EA. For this, each violation of a solution is weighted by 1000. Both the number of vias and the crosstalk effect are weighted by 10. The net length is weighted by 1 only.

In the following one fixed set of parameters is considered for all experiments. Notice, that in other applications different parameter settings might be of interest. Thus, the designer himself can influence the priorities of the optimization procedure.

Experimental Results

The method described above has been implemented in *C++*. All experiments have been carried out on a *Sun Sparc 20* workstation. In the experiments channels from [91] with up to 130 columns and 65 nets are considered. The runtimes for the examples considered range from several minutes (*ex3a* using two layers) up to one day (*ex5* using four layers) of CPU time.

In a first series of experiments the efficiency of the approach concerning the quality of the results is shown. The results are compared to [106] where a comparison to other routing approaches is given. As can easily be seen in

name	*columns*		EA	LT
Jo6_16	11	*rows*	6	6
		net length	115	116
		vias	16	15
		time	43.1 min	48.9 min
Bursteins Difficult Channel	12	*rows*	4	4
		net length	82	82
		vias	8	8
		time	0.5 min	9.6 min

Table 6.10 Results for small channels

name	*columns*		EA	EA*
fdc01	40	*rows*	4	4
		vias	18	16
		time	3.3 min	10.3 min
lprim1s	149	*rows*	50	50
		vias	104	117
		time	45.2 min	244.2 min

Table 6.11 Results for larger channels

Table 6.10 the quality of the routing solutions (column EA) is not worse than in [106] (column LT), but the runtime is improved.

To give an impression on the quality of the EA and especially of the initialization algorithm INIT the EA is applied to larger examples that are given in Table 6.11. *Iprim1s* is used in an industrial application and *fdc01* is a difficult channel that has been constructed for the application. Column EA shows the results obtained by the method described above. EA* denotes an EA where the initial population is obtained at random, i.e. without using INIT. As can be seen in Table 6.11 the application of the heuristic during the initialization directly improves the execution time, while the quality of the results are nearly the same.

name	*columns*	*nets*		2-layer	3-layer
			rows	10	8
ex3a	89	44	*vias*	99	162
			net length	1519	1418
			rows	12	9
ex3b	84	47	*vias*	131	178
			net length	1602	1562
			rows	12	9
ex3c	103	54	*vias*	132	181
			net length	2216	1809
			rows	13	9
ex4b	119	54	*vias*	184	190
			net length	2268	1876
			rows	14	10
ex5	129	65	*vias*	222	224
			net length	2575	2196

Table 6.12 Two-layer routing versus three-layer routing

In a next series of experiments channels with two layers and with three layers are considered. The routing solutions for two layers are compared to the solutions resulting from three-layer routing. This gives the reader who is not familiar with channel routing an impression how many tracks can be saved by using one additional layer. As can be seen in Table 6.12 the number of tracks that have to be used can significantly be reduced if additional layers are available. Furthermore, using three layers the net length is decreased. This directly corresponds to an improved delay behaviour of the resulting circuit.

In the next experiment realizations are considered that allow four layers. To give an idea on the quality of the results the EA is compared to the approach from [62] which is denoted by FFL in the following. In Table 6.13 it can be seen that the approach uses less tracks than FFL. Thus, the routing area could be significantly reduced. The EA approach needs more execution time, but the results are of higher quality. But using tracks for routing can directly result in lower fabrication costs, if the number of layers is fixed in a given technology. Additionally, less net length is needed. Only the number of vias is increased[5].

[5]In [62] a slightly simpler routing model has been considered, i.e. *surface connection* was not used. Thus, the number of vias obviously could be decreased.

name	*columns*	*nets*		FFL	EA
ex3a	89	44	*rows*	8	8
			vias	68	186
			net length	868	1352
ex3b	84	47	*rows*	9	7
			vias	107	198
			net length	1363	1457
ex3c	103	54	*rows*	9	7
			vias	125	254
			net length	1727	1871
ex4b	119	54	*rows*	9	8
			vias	179	257
			net length	2205	2031
ex5	129	65	*rows*	10	7
			vias	150	263
			net length	2200	2022

Table 6.13 Four-layer routing

Finally, we want to give an impression on the convergence behaviour of the optimization procedure using the objective function with parameter settings as given above. In Figure 6.17 the fitness values of the best element (with respect to the number of violations, vias, and net length) in a population are shown on the y-axis for channel *ex5* using four-layer realization. The x-axis gives the number of generations. (Notice, that due to the weights of the objective function the number of violations, vias, and net length given on the y-axis are not equal to the corresponding fitness.) It can be seen that first the number of violations, which are mainly penalized by the objective function, are reduced until the best element has no further violation. Then, the remaining attributes of the routing solution, like the number of vias and the net length, are optimized.

Conclusions

An EA using domain specific knowledge for multi-layer detailed routing has been presented. The main characteristics are the following:

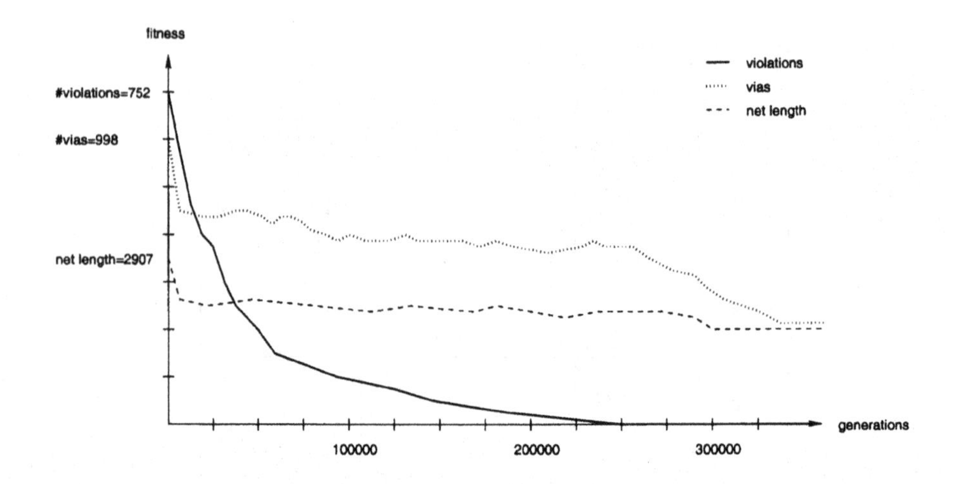

Figure 6.17 Behaviour of convergence of example *ex5*

- A variable size representation is used. By this the memory requirement is reduced, if the corresponding element of the population is optimized with respect to netlength. In contrast to other EA approaches this enables the possibility to also deal with larger problem instances.

- The operators of the EA are modified such that only cut positions are allowed between complete nets.

- The EA allows also invalid solutions. The validation is handled by the fitness function. Since in this application a repair run might become very time consuming this technique has shown to be very useful.

- Experiments have demonstrated the quality of the approach. Due to the fast execution times and its space efficient storing it can be applied to large problem instances, where other EA approaches fail. A comparison to previously published approaches has been provided to demonstrate the quality of the EA method. But again the EA is slower than standard rip-up and reroute techniques, that are here incorporated in the form of operators.

A possible extension of the approach presented here is to improve the considered model with respect to routing of *Multi Chip Modules* (MCMs). There the routing region can be modeled by channels, switchboxes, and obstacles. One main goal is then to find feasible routing solutions using a minimum number of layers. Recent extensions also consider parallel implementations [104, 105].

6.4 TESTING

Finally, in this chapter we focus on testing of circuits. We have a closer look at two different approaches to guarantee high quality tests:

1. First, an EA based method for test pattern generation is presented, i.e. for a given circuit and a fault (with respect to a given fault model) the algorithms tries to find a test or a test sequence, respectively.
2. Then an EA approach to design a *Linear Feedback Shift Register* (LFSR) is studied that can be used for build-in self-test.

6.4.1 Test Pattern Generation

Even if circuits are correctly designed, a non negligible fraction of them (sometimes up to 70%) will behave faulty because of physical defects caused by imperfections during the manufacturing process. Therefore, the detection of faulty circuits in a *test phase* is very important. *Automatic Test Pattern Generation* (ATPG), i.e. the construction of input vectors that detect all (or a high percentage) of faults (in a given fault model) is one of the key problems that has to be solved during this phase. This problem is known to be particularly hard for synchronous sequential circuits without reset line, i.e. circuits containing memory elements whose initial values are unknown. In this case a test consists of a finite sequence of test vectors $t_1, t_2, \ldots, t_m$ applied to the circuit in subsequent time frames starting with t_1 in time frame 1.

Two methodologies have been primarily used, i.e. deterministic algorithms and simulation based algorithms:

- State-of-the-art *deterministic fault oriented algorithms* (see e.g. [118]) for this problem are highly complex and time consuming, while
- *simulation based test generators* are fast but tend to generate longer test sequences and sometimes achieve lower fault coverage [2].

Recently, EAs have been proposed to overcome these difficulties. Before the approach from [77] is described in detail, we briefly comment on two other successful approaches in this area, i.e. [134, 135], to give an impression on the usefulness of of using EAs. Furthermore, this simplifies the understanding of different methods presented so far. Notice that evaluation of the fitness function in most EA approaches in mainly based on fault simulation.

In [134] test vectors with one bit per primary input are generated as initial population. An EA is applied to generate test vectors for subsequent time frames as long as "progress", i.e. an increase in fault coverage, can be observed. Then the resulting test sequence is enlarged by test sequences of fixed small length, which again have been determined by an EA. For the encoding of the candidate test sequences a binary or nonbinary coding can be used. (In a non-binary coding, each possible test vector is a separate character in the alphabet and the test sequence is viewed as a string over this alphabet.) Sequential fault simulation for the candidate elements is used to determine their fitness which takes into account the relative number of flip flops initialized, and the faults detected or propagated. Notice that by taking the flip flops into account also problem specific knowledge is incorporated in the fitness function. The fitness computation can be speeded up by restriction to a small sample of the faults and by use of overlapping populations. A large number of experiments is performed to determine the influence of different EA parameters on the performance of the algorithm: The progress limit for test vector generation and the length of the candidate test sequences is chosen as a small multiple of the sequential depth of the circuits considered. The number of generations is limited to 8 to reduce runtime. Among several selection schemes tournament selection without replacement turns out to be the most successful. Uniform crossover gives consistently better results than 1-point or 2-point crossover. Variations in mutation rate have only a small effect on fault coverage. And non-overlapping populations give the highest fault coverages.

Taken together, the experiments for the ISCAS89 benchmarks [15] demonstrate the effectiveness of this approach: In 13 of 17 cases the number of detected faults, i.e. the quality of the test sequence obtained, is comparable or better than for the HITEC deterministic fault-oriented test generator [118]. Test sequence length is reduced to 42% on average. In most cases test generation time is significantly smaller than for HITEC.

A disadvantage of the EA based method discussed before and of any simulation based method as well, is that redundant or untestable faults cannot be classified. To overcome this problem, EA based techniques are combined with deterministic techniques in [135], the later being applied for the identification of redundant and untestable faults and for the correction of the EA when a fixed number of test vectors was added without improving the fault coverage. In principle, the methods for the EA part are similar to the ones proposed in [134], though the fitness function is different: It is based on balanced circuit activity as recorded by a logic simulator (giving equal chance to every fault to be activated). Only when a few undetected faults are left, the extent of fault propagation to primary outputs is checked. The proposed techniques are applied to several ISCAS89 benchmarks [15] and run on the average 10 times faster than traditional deterministic techniques. Fault coverage and test length are comparable.

Also in this area several extensions of the basic principles have been reported [28, 30, 84, 125]. The methods differ with respect to integration of deterministic ATPG tools, in choice of operators, resulting test length.

In the following a (relatively) simple EA is described that has the following main features:

- It does not make (intensive) use of problem specific knowledge, e.g. the number of flip flops already initialized is not considered.
- The EA uses a fault simulator with a high accuracy, i.e. instead of a three-valued simulator a symbolic fault simulator is used. (Symbolic methods have before only be used for deterministic approaches [22].) The symbolic fault simulator used is the one from [97] based on OBDDs. By this it is more accurate than the classical fault simulators based on three-valued logic used in the approaches outlined above. Unfortunately, OBDDs can not be constructed for all functions, since too many OBDD nodes are needed. For this, the simulator is able to switch between several simulation modi, i.e. three-valued, symbolic, and mixed. As long as sufficient memory is available it performs symbolic simulations. If a given memory limit is reached it dynamically switches to a less accurate mode. After some simulation steps it returns to the symbolic mode and in this sense it can be seen as a *hybrid* method. Depending on the space requirement during simulation the hybrid fault simulator chooses the most powerful simulation type available.
- The ATPG algorithm consists of two phases:
    1. A test sequence is optimized with respect to fault coverage and test length using an EA for a given sequence of time frames.

2. The test sequence is enlarged to improve the fault coverage.

- Since a more time consuming and more accurate fault simulation procedure is used, additionally the number of evaluations (fault simulations) is reduced during the EA run. This is done by an alternative selection strategy which allows to evaluate only "essential" individuals.

- The EA makes use of a new genetic operator that includes all previously presented operators for EA based ATPG. By this, a unified framework is described.

All in all, the EA makes use of several "EA specific" tuning techniques, but no combination with deterministic approaches is used. Its main feature is the symbolic fault simulator. Thus, the approach is fully compatible with other approaches making use of deterministic algorithms. To simplify the understanding we restrict in the following to this simpler approach.

Preliminaries

First, the notations and some basic definitions are introduced. Synchronous sequential circuits are defined and the underlying fault model is described. Then the problem domain is formulated. Finally, the main featured of the fault simulator used are reviewed.

A synchronous sequential design can be described using a *Finite State Machine* (FSM):

- A FSM is a 5-tuple $M = (I, O, S, \delta, \lambda)$, where I is the input set, O is the output set and S is the set of states. $\delta : I \times S \rightarrow S$ is the next-state function, and $\lambda : I \times S \rightarrow O$ is the output function.

- Since in the following a gate level realization of the FSM is considered, we have $I = \mathbf{B}^k$, $O = \mathbf{B}^h$ and $S = \mathbf{B}^m$ with $\mathbf{B} = \{0, 1\}$. k denotes the number of primary inputs, h denotes the number of primary outputs, and m denotes the number of memory elements. The functions δ and λ are computed by a combinational circuit C.

- The inputs (outputs) of the combinational circuit, which are connected to the outputs (inputs) of the memory elements, are called secondary inputs (outputs). The secondary inputs are also called ***present state variables*** and the secondary outputs are called ***next state variables***.

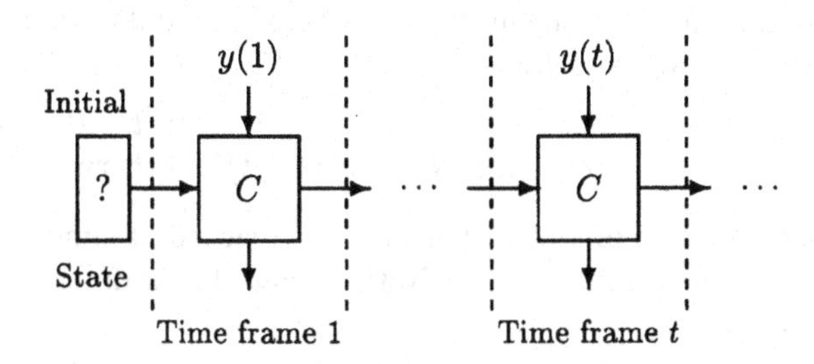

Figure 6.18 Iterative description of a sequential circuit

For the description of the algorithms the following notations are used:

- $Y = y(1), \ldots, y(n)$ denotes an input sequence of length n and $y_i(t)$, $1 \leq i \leq k$, denotes the value that is assigned to the ith primary input before starting simulation at time t, $1 \leq t \leq n$.

- $S(Y) = s(0), s(1), \ldots, s(n)$ denotes the state sequence defined by $s(0)$ and Y. $s_i(t)$, $1 \leq i \leq m$, denotes the state of the ith memory element after simulation step t. $s(0) = s_0$, $s_0 \in S$, is called the initial state.

- $O(s_0, Y) = o(1), o(2), \ldots, o(n)$ denotes the output sequence defined by the initial state s_0 and Y. $o_i(t)$, $1 \leq i \leq h$, denotes the value at the ith primary output after simulation step t. Using these notations the next state is given by

$$s(t) = \begin{cases} s_0 & \text{if } t = 0 \\ \delta(y(t),\, s(t-1)) & \text{otherwise} \end{cases}$$

  Analogously, the output $o(t)$ is defined by the function λ.

Thus, a synchronous sequential design is considered as an iterative network (see Figure 6.18).

Remark 6.2 In the following an unknown initial state is assumed, i.e. we do not presume a given reset state.

To describe the simulation of the machine starting in an unknown initial state, a function *s_set* is defined as follows:

$$s\_set(t) = \begin{cases} \mathbf{B}^m & \text{if } t = 0 \\ \{\delta(y(t), s) | s \in s\_set(t-1)\} & \text{otherwise} \end{cases}$$

$s\_set(t)$ represents the set of states which can be reached at time t, $1 \leq t \leq n$, starting in an unknown initial state. (Notice, that this is different to several other ATPG approaches.)

It is well-known that, even if the circuits are correctly designed, a fraction of them will behave faulty because of physical defects caused by imperfections during the manufacturing process. Fault models, which cover a wide range of the possible defects, have to be defined, and tests for faults in the fault models have to be constructed. The fault model adopted in this book is described in the following.

The *Stuck-At Fault Model* (SAFM) [51, 14] is well-known and used throughout the industry. It verifies the logical behaviour of the circuit. The single-fault assumption is used, i.e. it is assumed that at most one fault can occur in the circuit.

Definition 6.1 A *stuck-at fault* f transforms a machine M into a machine $M^f = (I, O, S, \delta^f, \lambda^f)$. The functions δ^f, λ^f and s_set^f are defined analogously. Notice, $s\_set(0) = s\_set^f(0)$ for any fault.

If there is no knowledge about the initial state of the machine the detectability of a stuck-at fault with respect to an input sequence Y can be defined according to [1] as follows:

Definition 6.2 : A fault f is *detectable* by a sequence $y(1), y(2), \ldots, y(n)$ if

$\exists t \leq n$, $\exists i \leq h$, $\exists b \in \mathbf{B}$ such that

$$\forall\, s \in s\_set(t-1)\colon\ o_i(t) = b \text{ with } o(t) = \lambda(y(t), s)$$

and

$$\forall s \in s\_set^f(t-1)\colon\ o_i^f(t) = \bar{b} \text{ with } o^f(t) = \lambda^f(y(t), s).$$

For combinational circuits, each detectable fault can be determined by a single vector, i.e. a test vector that detects the fault. This is no longer the case for sequential circuits, in general. There, a sequence of input vectors, i.e. a test sequence, is used to detect a certain fault. The "classical" test pattern generation problem for sequential circuits is formulated as:

How can a test sequence be determined such that a given fault f is detected by this sequence, starting from an unknown initial state ?

As outlined above in the approach described here a hybrid fault simulator is used. Basically, the purpose of fault simulation is to determine, which faults are detected by a given sequence of input values, according to Definition 6.2. The number of detected faults, divided by the total number of faults is named *fault coverage.*

A common strategy for handling the problems induced by an unknown initial state is given by a fault simulation procedure based upon a three-valued logic over $\mathbf{B}_X := \{0, 1, X\}$. X denotes the unknown or undefined value. It is used for encoding the unknown initial state, which is given by $s_0 = (X, X, \ldots, X)$.

As is well-known, fault simulation based upon the three-valued logic is rather inaccurate. Many papers mention this problem (see e.g. [23, 113]). The inaccuracy can be explained on the one hand by the binary encoding of the states [113] and on the other hand by the logic realizing the next state function. Thus, there may exist detectable faults, that cannot be detected with respect to a given input sequence Y using fault simulation based upon the three-valued logic. Therefore, these fault simulators only determine a lower bound for the "real" fault coverage.

A lot of work has been done to overcome these problems. On the one hand, a full or partial scan design can help [3, 21]. But a scan design involves additional circuitry such as multiplexers, which increase the area and degrade the performance of the circuit [3]. On the other hand, there is the application of symbolic traversal techniques to test generation [23, 147]. This approach is based upon *Ordered Binary Decision Diagrams* (OBDDs) [18]. Using symbolic methods, the exact fault coverage with respect to a given input sequence Y according to Definition 6.2 is determined.

In the following the event-driven *Hybrid Fault Simulator* (HFSim) from [97] is used and its main features are briefly reviewed[6]:

[6]In this context *hybrid* means, that the algorithm supports different logics for simulation and allows a dynamic switching to the logic used for each simulation step.

- HFSim consists of three kinds of fault simulation procedures:

  (1) a fault simulation procedure based upon the three-valued logic,

  (2) a symbolic fault simulation procedure based upon OBDDs, and

  (3) a fault simulation procedure, which is hybrid itself, in the sense that a symbolic true value simulation and an explicit fault simulation procedure based upon the three-valued logic are combined.

  The three fault simulation procedures differ in their time and space requirements and the accuracy of the fault coverage.

- The fault simulation tool chooses automatically the best fault simulation procedure depending on time and space requirements of the simulation and the user's parameter settings at the beginning.

Formally, the symbolic (fault) simulation is based upon the logic

$$\mathbf{B}_m = \{g : \mathbf{B}^m \rightarrow \mathbf{B}\}.$$

For the simulation, an element of the logic is represented by an OBDD. The two constant functions **0** and **1** contained in $\mathbf{B}_m$ are used for encoding the elements of the input vector. To encode the unknown initial state a Boolean variable x_i, $1 \leq i \leq m$, is introduced for each memory element representing its unknown value at the beginning of the simulation. The initial state of the good machine as well as of the faulty machine is given by $s_0 = (x_1, x_2, \ldots, x_m)$. A fault f is marked as detectable according to Definition 6.2, if there exists an output for which M and M^f compute different constant functions.

HFSim tries to combine the advantages of these procedures by choosing a convenient logic before starting the simulation for the next input vector. For that reason, the hybrid fault simulation strategy determines the exact fault coverage or at least a tighter lower bound than the three-valued approach. (For more details see [97].)

Evolutionary Algorithm

Now the EA that is used for generating test sequences for sequential circuits is described. After briefly explaining the overall flow of the algorithm the representation is described. Then some more details on the evaluation of the fitness function (that is mainly based on fault simulation as described above) are given. A sketch of the algorithm is given and the parameter settings are discussed.

To simplify the understanding a short overview of the EA is given:

- The environment for ATPG makes use of two main EA phases:
  (1) A first phase optimizes test sequences to yield high fault coverage.
  (2) In the second phase the EA enlarges this optimized test sequence to improve the results.
- The approach is based on new genetic operators and the hybrid fault simulator. Additionally, the performance with respect to runtime is reduced while using an other selection process which reduces the number of simulations tremendously.
- By this combination high fault coverages are obtained. As a "side effect" the final test sequences are often very short in comparison to other approaches (see below).

Representation

First, the representation of test sequences in a population $\mathcal{P}$ is described. An individual is a binary string of length l that represents a test sequence Y of n time frames. It is clear that $n = l/k$, where k denotes the number of primary inputs. The maximum length l of the individuals is fixed, i.e. each test sequence in a population has at most length l. The choice of n depends on the present phase of the algorithm and the size of the circuit being considered.

Objective Function and Selection

The objective function computes the fitness for each element in the present population. It measures the number of undetected faults and the length of the test sequence. Thus, the objective function has to be minimized. Notice, that the primary optimization goal is the improvement of the fault coverage. The second optimization criterion is the minimization of length of the test sequences. The elements in a population are evaluated by the objective function, that basically consists of fault simulation.

The selection of the elements is done by *2-tournament selection*:

2-tournament selection: For each parent two elements are chosen randomly. Only these two elements are then evaluated by the objective function. The element with higher ranking is then selected. (Notice, that by this for the selection of two parents only four elements are evaluated instead of the whole population. For the population sizes used in the following this results in a speed-up of more than a factor of 5 in comparison to roulette wheel selection.)

Genetic Operators

First, it is distinguished between "standard" crossover operators and problem specific operators (see e.g. [30, 28]). In the following only the standard operators are used and one new operator that is a generalization of others presented so far. Additionally, "classical" mutation operators are applied to guarantee the diversity in the population. All operators are directly applied to binary strings of length l that represent elements in the population. The parent(s) for each operation is (are) determined as described above. First, the standard EA operators are used, i.e. crossover, 2-time crossover, and uniform crossover.

Next, the problem specific operator is presented. The string representation of a test sequence is interpreted as a two-dimensional matrix, where the x-dimension represents the number of inputs and the y-dimension represents the number of test pattern. The operator works as follows:

Free Vertical Crossover: Construct two new elements x_1 and x_2 from two parents y_1 and y_2. Determine for each test vector t a cut position i_t. Divide each test vector t of y_1 and y_2 in two parts at cut position i_t. The first (second) part of each test vector of x_1 (x_2) is taken from y_1 and the second (first) part is taken from y_2. (Notice, that the *vertical crossover* from [30] is a special case of this operator, if i_t is equal for all test vectors t.)

Example 6.8 The behaviour of the *free vertical crossover* is illustrated in Figure 6.19. The black filled areas result from the different length of the sequences; they are filled with random pattern.

Moreover, three (standard) mutation operators are applied which are based on bit-flipping at a random position, i.e. mutation, 2-time mutation, and mutation with neighbour.

Remark 6.3 All genetic operators generate only valid solutions, if they are applied to binary strings.

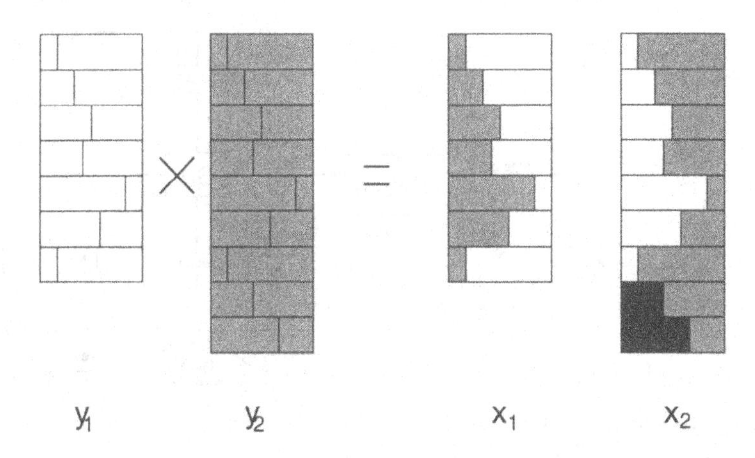

Figure 6.19 Free vertical crossover

Algorithm and Parameter Setting

Now the basic EA is introduced that is the kernel of the optimization procedures. It is used in both phases of the ATPG algorithm.

- The initial population of size $|\mathcal{P}|$ is generated, i.e. the binary strings of length l are initialized using random values.
- Two parent elements are determined.
- Two new individuals are created using the genetic operators with given probabilities.
- These new individuals are then mutated by one of the mutation operators with a fixed mutation rate.
- The two elements which loose the tournament selection in the present parent population are deleted and the offsprings are inserted in the population.
- The algorithm stops if the best element has not changed for a fixed number of generations.

As mentioned above the EA based ATPG algorithm consists of two phases. For illustration a structure of the algorithm is given in Figure 6.20.

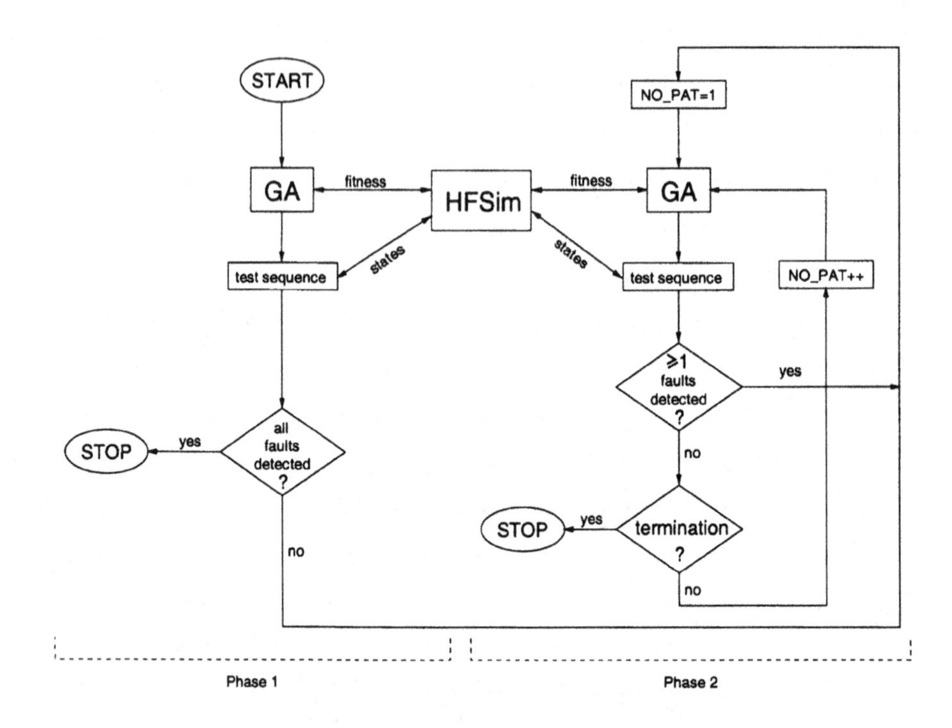

Figure 6.20 Structure of the algorithm

Phase 1: The algorithm starts optimizing a test sequence of fixed length beginning in the unknown initial state. This test sequence is optimized with respect to fault coverage and length of the test sequence. The initial population is generated, where max_pat is the number of test vectors. max_pat depends on the circuit considered.

HFSim starts in the initial state $s_0 = (x_1, \ldots, x_m)$. The test sequence that results from Phase 1 is represented by the best element and is denoted by Y; $|Y|$ denotes the length of the test sequence. The present state of HFSim after Phase 1 is $s(|Y|)$ which is the starting state for Phase 2. A sketch of Phase 1 is given in Figure 6.21.

Phase 2: Tests are generated for the remaining undetected faults, i.e. the test sequence Y from Phase 1 is enlarged by the test vectors (test sequences) that have been created. For this, the EA does not start in the unknown initial state, but in state $s(|Y|)$ that results from applying Y.

```
phase_1 (circuit):
    generate_initial_population (max_pat) ;
    calculate_fitness ;
    do
        select_parents ;
        apply_operators_with_corresponding_probabilities ;
        calculate_fitness ;
        update_population ;
        while (not terminal case) ;
    Y = best_element ;
    return Y;
```

Figure 6.21 Sketch of Phase 1

The EA's second phase is started using no_pat = start_pat test vectors. (See illustration of Phase 2 in Figure 6.22.) If the EA terminates successfully, i.e. a test vector has been found, this test vector is suffixed at the end of Y. This newly generated test sequence composed of Y and the new test vector is denoted by Y_{new}. If no fault has been detected, the length no_pat of the additional test sequence is incremented by offset, otherwise it is set to no_pat = start_pat. Then the EA of the second phase is restarted in the new present state $s(|Y_{new}|)$. This is repeated, until no further fault is found for a given upper bound.

The size of the population in Phase 1 is set to $|\mathcal{P}| = 32$ and in Phase 2 it is set to $|\mathcal{P}| = 16$. If Phase 2 runs in the three-valued mode the population size is enlarged to $|\mathcal{P}| = 32$, since the three-valued simulation runs much faster and thus more time can be spend for different evaluations of the EA. Each recombination operator is performed with a probability of 20% and the mutation rate is set to 5%. If no improvement is obtained for 100 generations the EAs stop.

Experimental Results

The ATPG algorithm described above is applied to several sequential benchmark circuits from ISCAS89 [15]. The algorithm is compared to an other EA

```
phase_2 (circuit):
    no_pat = start_pat ;
    Y_new := Y ;
    do
        generate_initial_population (no_pat) ;
        calculate_fitness ;
        do
            select_parents ;
            apply_operators_with_corresponding_probabilities ;
            calculate_fitness ;
            update_population ;
            while (improvement obtained) ;
            if (fault(s) detected)
                Y_new := Y_new ∘ best_element ;
                no_pat = start_pat ;
            else
                no_pat = no_pat + offset ;
        while (not terminal case) ;
    return Y_new ;
```

Figure 6.22 Sketch of Phase 2

based method from [134][7]. Furthermore, the EA methods are compared to the deterministic ATPG tool HITEC [118]. Notice once more that the different approaches compared in the following make use of completely different techniques to improve the fault coverage.

The results are given in Table 6.14. The name of the circuit is given in the first column followed by the number of primary inputs and secondary inputs in columns two and three, respectively. For all ATPG tools the fault coverage FC and the test length $|v|$ obtained is reported. The values for HITEC are given first, since this is the "classical" ATPG method. As can be seen by a comparison with the EA approaches HITEC computes very long test sequences, i.e. up to a factor of 10 longer, while achiving in some cases lower fault coverages (see e.g. *s526*). The column denoted by EA_{RPGN} shows the fault coverages and

[7] Comparisons to [28, 30] are not given, since there tests are generated for circuits where a reset state is assumed while in the approach here considers the harder problem of circuits without a reset state or a reset sequence.

name	PI	SI	HITEC		EA_{RPGN}		EA_{X01}		EA_{HFSim}	
			FC	$\|v\|$	FC	$\|v\|$	FC	$\|v\|$	FC	$\|v\|$
s298	3	14	86.04	306	86.04	161	86.04	123	87.34	129
s344	9	15	95.91	142	96.20	95	96.20	69	97.37	91
s349	9	15	95.71	137	95.71	95	95.71	60	96.86	58
s386	7	6	81.77	311	76.82	154	78.39	1196	74.22	74
s400	3	21	90.33	4309	85.68	280	82.78	268	58.73	43
s510	19	6	–	–	–	–	0.00	–	100.00	402
s526	3	21	66.00	2232	74.96	281	75.32	598	67.00	166
s641	35	19	86.51	216	86.51	139	86.51	117	87.37	326
s713	35	19	81.93	194	81.93	128	81.93	114	82.27	83
s820	18	5	95.64	984	60.71	146	49.23	145	53.06	1567
s832	18	5	93.91	981	61.95	150	56.60	145	60.69	1003
s1196	14	18	99.20	453	99.19	347	99.19	372	99.76	775
s1238	14	18	94.69	478	94.02	383	94.02	353	94.61	838
s1488	8	6	97.17	1294	93.67	243	91.45	324	91.39	682
s1494	8	6	96.48	1407	94.02	245	91.43	1061	91.57	664
s5378	35	179	70.35	941	68.98	511	69.65	834	70.15	1030
s35932	35	1728	89.27	439	89.55	197	82.57	100	89.22	144

Table 6.14 ATPG using three-valued simulation

number of test vectors that were obtained in [134] and the column denoted by EA_{X01} gives the results obtained by the method described above, if restricted to a pure three-valued simulation, i.e. the symbolic techniques are not used. In column EA_{HFSim} the results using the hybrid approach are given. As can be seen the three-valued EA approach EA_{X01} obtains about the same fault coverages as EA_{RPGN}, but the test sequences are often shorter. Due to the higher accuracy of the symbolic techniques the best known fault coverages can often be further improved. This can range up to some orders of magnitude, e.g. for circuit *s510* not a single fault can be detected using a three-valued simulator, while the OBDD based approach obtains a fault coverage of 100%.

On the other hand for larger circuits (where also the search space becomes much larger) the "pure" simulation based technique described here does not obtain results of the same quality as the deterministic tool HITEC.

Conclusions

An EA for *Automatic Test Pattern Generation* (ATPG) for sequential circuits has been presented. The main characteristics of EA based ATPG are:

- The EA based methods obtain often high quality results, but in contrast to many other applications in VLSI CAD also improvements with respect to runtime have been reported by many authors. These positive effects result from a clever combination of deterministic and simulation based approaches.
- Motivated by these positive observations EAs have been proposed that mainly vary by the following aspects:
  - type of operators
  - three valued simulation vs. symbolic simulation (see above)
  - pure simulation vs. combination with deterministic approach
- Using hybrid symbolic fault simulation has several advantages concerning the quality of the resulting fault coverages. But also the runtime needed for fault simulation increases. If complete symbolic evaluation is no longer possible due to memory requirements induced by the OBDDs the fault simulator can switch dynamically to a less accurate simulation mode. By this technique the approach can also handle large circuits.
- The results showed that for some circuits the "pure" EA approaches fail, while deterministic algorithm work quite well. This again underlines that integration of problem specific knowledge turns out to be (at least) as important as high accuracy during simulation.
- Furthermore, by the combination of both approaches also the runtime of the EA can be reduced. One major problem of EAs when applied to ATPG is that they can not detect redundant faults. Thus, they often waiste a lot of time trying to generate a test, while non exists. At this point a deterministic algorithm can be combined with the EA approach quite well.

Recently, several extensions of the basic ideas discussed above have been proposed (see [28, 30, 84, 125]).

6.4.2 Build-In-Self-Test

Recently, EAs have also been applied in other testing areas, like e.g. partial scan flip flop selection [29]. In the following an approach to build-in-self-test from [121] is described.

A built-in test based on pseudo random patterns generated by a *Linear Feedback Shift Register* (LFSR) is an alternative to external testing. Often the fault coverage achieved by a random test sequence is too low or the length of the test sequence which leads to an acceptable fault coverage is too large. One approach to improve the quality of built-in self test schemes is the optimization of input probabilities for *Weighted Random Pattern Generation* (WRPG). There is a trade-off that has to be optimized:

- A small number of weight vectors implies a large test application time.
- A small test application time requires a large number of weight vectors, leading to a high area overhead.

Different methods for computing weight vectors by optimizing fault detection probabilities have been presented [5, 81, 88, 126]. Since computing fault detection probabilities is NP-hard, these methods are based on approximated values [100, 143, 157] which may be inaccurate and thus misdirect the optimization process. Instead here a numerical optimization technique based on exact fault detection probabilities computed by evaluating the OBDD of the fault detection function is used (see [98]).

Experimental analysis showns the applicability of the approach and the superior quality of weight vectors based on continuous (exact) weight values. But if the random test is to be implemented as a BIST scheme, the weighted random patterns must be determined by rounding. However, this may lead to a significant deterioration of the quality of the results.

These disadvantages can be avoided by using EAs instead of numerical methods:

- Although EAs generally will not yield global optima as well, they turn out to be the better heuristics. In particular, EAs seem to be predestined for discrete optimization as they do not demand the objective functions to have special properties, like continuity or differentiability. This merit of EAs allows to use other than the classical objective functions for improving random BIST.

- Optimizing classical objective functions, e.g. those introduced in [81, 158], aims at the statistical improvement of random testability, but will not always provide satisfactory results concerning the actual test application time on-chip. Therefore an objective function is proposed which measures the quality of the BIST structure. This quality is given by the actual test length or fault coverage, using fault simulation with weighted random patterns generated by the BIST structure.

Thus, the optimizing strategy is as follows:

- During the first stage of the EA, discrete input probabilities are optimized using the objective function presented in [81].
- During the second stage, all other parameters affecting the BIST structure, like feedback connections, will be optimized.
- The output of the EA will not just consist of optimized input probabilities, but the complete hardware design of the BIST structure.

Experimental results show, that the optimized, discrete weights are as good as the exact, continuous weights, concerning test length and fault efficiency. Moreover, using a space efficient BIST realization based on ideas of [82] optimized by the EA's second stage, a fault efficiency of at least 99.95% is achived with only two weight vectors for all circuits considered.

Preliminaries

Analogously to the last section the gate-level realization of a combinational circuit C and the (single) stuck-at fault model is used.

Definition 6.3 Let F denote the set of all stuck-at faults of C. The *detection function of a fault* f is given by $T_f : \mathbf{B}^n \to \mathbf{B}$ with $T_f(x) = 1$ iff x is a test for f, where n denotes the number of primary inputs of C.

Some further notations are needed in the following:

- Let $g : \mathbf{B}^n \to \mathbf{B}$ be a Boolean function depending on Boolean variables $x_1, \ldots, x_k$. These variables are interpreted as independent random variables defined on the probability space $\Omega = \mathbf{B}^n$.

- The probability distribution that defines the probability of such a random variable assuming the value 1 is denoted by P_r. These probabilities are abbreviated by $q_i = P_r(x_i = 1)$ and denote q_i, $1 \leq i \leq n$ as *input probability* or *weight value.* (The weight value is also called *weight* for short.)

- $\vec{q} = \{q_1, ..., q_n\}$ represents the so-called *weight vector.*
- $P_r(g = 1)$ denotes the probability that the Boolean function g assumes the value 1 for a random argument depending on P_r.

As is well-known, each Boolean function can be represented by an OBDD [18, 13]. If the OBDD representation of a Boolean function g is given, the probability $P_r(g = 1)$ can be determined in linear time [9, 95].

The detection probability of a fault f in a combinational circuit expresses the probability that f will be detected by a randomly selected input vector according to the probability distribution P_r. The detection probability of f is denoted as $\delta(f) = P_r(T_f = 1)$.

The OBDD representation of the fault detection function can efficiently be constructed [99].

- The representation of the fault detection functions as an OBDD allows only a recursive computation of the fault detection probabilities.
- An acceleration can be achieved by sorting the nodes of the OBDDs in a topological order and storing the nodes in a list. This has to be done separately for each subset only once.
- Afterwards, the computation of fault detection probabilities can be done iteratively by traversing the lists. This iterative computation is considerably faster than recursive computation.

In the following the basic underlying concepts of the EA approach are described. We start with the definition of discrete weight sets. Furthermore, the target hardware realization from [82] is introduced. It turns out, that a given weight vector does not define this BIST structure in all aspects. For example the characteristic polynomial of the LFSR is not specified. The remaining information is computed using an EA, which is introduced below.

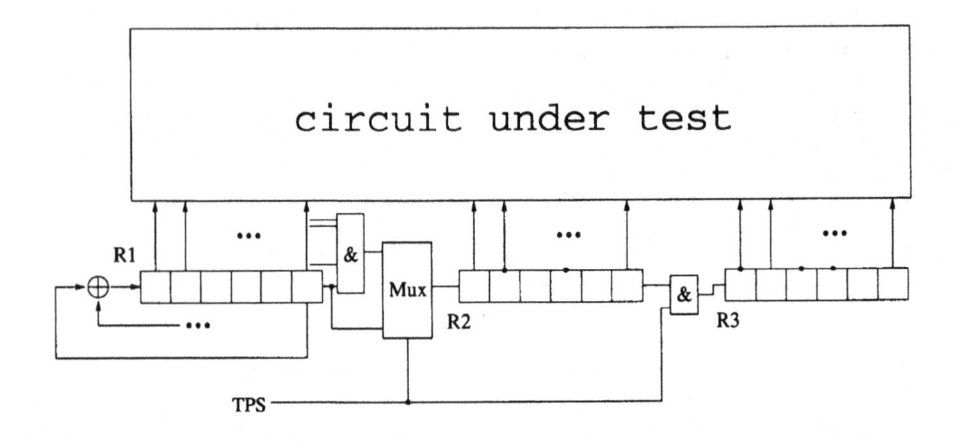

Figure 6.23 BIST structure of [82]

If a random test is to be implemented as a BIST scheme, the random patterns must not be based on exact (continuous) weights. The hardware design of most of the LFSR based random pattern generators restricts the weights that can be used to those contained in the following set:

$$W_{a,b} := \{0, \frac{1}{2^{a+b}}, \frac{1}{2^{a-1+b}}, \ldots, \frac{1}{2^{2+b}}, \frac{1}{2}, \\ 1 - \frac{1}{2^{2+b}}, 1 - \frac{1}{2^{3+b}}, \ldots, 1 - \frac{1}{2^{a+b}}, 1\}$$

This set is denoted as (a, b)-valued discrete weight set, where $a \geq 2$ and $b \geq 0$. Obviously, the definition implies $|W_{a,b}| = 2 \cdot a + 1$.

In [155, 82] procedures for weight optimization are presented which take into account these restriction. The results show that the use of discrete weights usually reduces the quality of the continuous weight vectors in proportion to the number of weights permitted.

Using LFSRs [5] only unweighted pattern can be generated. If both unweighted and weighted random patterns are to be applied an additional circuitry for switching from unweighted to weighted random pattern generation is needed. In [82] a space efficient implementation with this property is introduced. Since this hardware is optimized by the EA, its structure is briefly described (see Figure 6.23):

- This BIST structure contains an LFSR R_1 of length l_1 and two ordinary shift registers R_2 and R_3 of lengths l_2 and l_3, respectively.
- The outputs of the elements in R_3 are connected to those primary inputs of the circuit which according to the given weight vector shall be supplied with constantly 0 or constantly 1 during weighted random pattern generation.
- Analogously, the elements in R_2 lead to primary inputs to be set to 1 with probability of $\frac{1}{2^{2+b}}$ or $1 - \frac{1}{2^{2+b}}$, respectively.
- Primary inputs connected with elements in R_1 will be set to 1 with uniformly distributed probability.

The BIST structure works as follows:

- During the first test phase *Test Phase Select* (TPS) is set to 0, causing R_3 to be provided with the constant value 0. R_2 is fed with the logical AND of values derived from EXOR gates, connected to storage elements $r_1, \ldots, r_{4+2b}$ in R_1. These $(2 + b)$ 2-input EXOR gates help to reduce correlations between elements of R_1 and R_2 [82].
- After applying a certain number of weighted random patterns TPS is switched to 1, causing all values at the output of LFSR R_1 to be shifted through R_2 and R_3. Note, that if there is already a shift register structure on chip, the hardware overhead reduces to a few basic gates.

The hardware is to a great degree determined by a given weight vector $\vec{q}$ with $q_i \in W_{2,b}$, $1 \leq i \leq n$. Therefore, it is referred to it as a $(2, b)$-valued WRPG. Nevertheless, results obtained performing a weighted random BIST often do not come up to expectations, even if the input probabilities they are based on are of high quality. This can be explained by problems related to the following questions:

Which polynomial should be selected for R_1?
Where should the connections be placed in R_1 for R_2?
How to initialize the BIST structure?

Often it is impossible to solve these problems efficiently by hand, so an algorithm that finds a good solution is strongly necessary. These problems get even worse, if only a small part of the maximum period (an m-sequence [5]) of an LFSR is used for WRPG. To overcome these problems an EA is used to search for solutions.

Evolutionary Algorithm

As outlined in Chapter 2 EAs represent an alternative to classical numeric optimization methods, like e.g. the well known gradient method, that demands an objective function that is differentiable and strictly convex [158]. Using discrete weight values will cause the objective function to lose the property of being continuous, so it will no longer be suitable for numeric optimization. In such cases probabilistical approaches like EAs are a very promising alternative. In the following a two-staged EA is presented for optimizing (a, b)-valued WRPGs with regard to the quality of the pseudo random tests they produce.

Representation

The information contained in an individual is given by:

Definition 6.4 With $q_j \in W_{2,b}$, $t_j \in \mathbf{B}$, $1 \leq j \leq n$, $v_j \in \{1, ..., l_1\}$, $1 \leq j \leq 4 + 2b$, and $r_j \in \{1, ..., l_1\}$, $1 \leq j \leq 2$, a $(2, b)$-valued individual is a vector

$$\vec{c} = q_1, ..., q_n, t_1, ..., t_n, v_1, ..., v_{4+2b}, r_1, r_2.$$

$\vec{c}$ consists of the following parts:

1. $\vec{q}$ represents a $(2, b)$-valued discrete weight vector.
2. $\vec{t}$ represents an initial state of a $(2, b)$-valued WRPG.
3. $\vec{v}$ denotes the storage elements whose outputs feed the EXOR/AND gates to perform weighting.
4. $\vec{r}$ denotes the storage elements which have a feedback connection.

Note that the design of a $(2, b)$-valued WRPG is uniquely determined by such a $(2, b)$-valued individual. In other words, not only the weight vector is encoded into an individual, but also the complete description of the target hardware. By this, the best individual at the end of the algorithm not only determines an optimized weight vector, but also a hardware realization of that vector.

To minimize hardware costs, $\vec{r}$ is restricted to a pair of storage elements. Hence, the characteristic polynomial associated with LFSR R_1 will be of the shape

$$f(x) = 1 + x^{r_1} + x^{r_2}$$

for a $(2, b)$-valued WRPGs defined by an $(2, b)$-valued individual. A set of N $(2, b)$-valued individuals is denoted as a $(2, b)$-valued population of size N.

Objective Function and Selection

Within the scope of the two-staged EA, it is intended to optimize individuals with respect to the minimization of two objective functions, i.e. $\varphi_r(\vec{c})$ and $\varphi_s(\vec{c})$. The fitness of an individual $\vec{c}$ denotes the value computed by $\varphi_r(\vec{c})$ or $\varphi_s(\vec{c})$.

1. During the first stage, individuals are optimized with respect to the expected test length $\bar{L}(\vec{c})$. $\bar{L}(\vec{c})$ is approximated by φ_r [81], with $r = 1$:

$$\bar{L}(\vec{c}) \approx \varphi_r(\vec{c}) := \sum_{f \in F} (\frac{1}{\delta(f)})^r.$$

   With that formula, only the weight vector $\vec{q}$ is affected by the optimization.

2. During the second stage, individuals are optimized with respect to the actual fault coverage yielded by weighted random pattern sequence of length L. A maximum number of pattern is set to $\tilde{L} = 2^{14} = 16384$ ($\tilde{L} = 5000$ for the combinational circuits). $\tilde{L}$ is doubled if no individual yielding complete fault coverage is created after a certain number of iterations. When $\tilde{F}$ denotes the set of detected faults, the fitness of each individual is computed by the objective function:

$$\varphi_s(\vec{c}) = L \cdot (2 - \frac{|\tilde{F}|}{|F|})$$

   If no faults remain undetected, $\varphi_s(\vec{c})$ equals the test length L. Otherwise $L = \tilde{L}$ is weighted with a (penalty) factor.

In particular, φ_s guarantees a further improvement even if all individuals yield a complete fault coverage. In this case optimization continues with respect to the test length.

In this application an alternative selection method is applied that is usually used for *evolution strategies* [128, 4] (see also Section 2.4):

The $(N + \lambda)$-method in combination with (N, μ)-selection is used. The λ best individuals of the preceding population are added to the current population. From the resulting set of $N + \lambda$ individuals only the best $\mu \leq N$ constitute a set of privileged individuals on which genetic operators may be performed. From this set pairs of individuals (parents) are chosen at random to be duplicated. With given probabilities the corresponding genetic operators (see below) are iteratively applied to the duplicates converting them into new individuals, i.e. the children of the population. An iteration step ends after N children have been created. The children then represent the new population. This selection method guarantees that the best individuals will never be lost to the reproduction process. (As can be seen the basic underlying selection principles are similar to classical EAs.)

Genetic Operators

During the first stage of the EA the operators are applied only on the weight vector. In the second stage the mutation operator may be applied to all vectors an individual consists of, whereas the crossover operator is restricted to the weight vector only. In this EA crossover and 2-time crossover is used.

The mutation proceeds as follows: With probabilities

$$\frac{P_{mut}(\vec{q})}{n}, \frac{P_{mut}(\vec{t})}{n} \text{ and } \frac{P_{mut}(\vec{v})}{2+b}$$

each component of the corresponding vector is changed to a value chosen at random from $W_{2,b}$, $\mathbf{B}$ and $\{1, ..., l_1\}$, respectively. Hence, $P_{mut}(\vec{q})$, $P_{mut}(\vec{t})$ and $P_{mut}(\vec{v})$ denote the expected number of components to be changed in the corresponding vectors of each chosen individual in each iteration step. We denote mutation with regard to $\vec{q}$, $\vec{t}$, $\vec{v}$ and $\vec{r}$ as $mut(\vec{q})$, $mut(\vec{t})$, $mut(\vec{v})$ and $mut(\vec{r})$, respectively. To avoid the generation of invalid individuals, with respect to the target hardware, it is guaranteed that $v_1, ..., v_{4+2b}, r_1, r_2 \leq l_1$.

Using the listing of primitive polynomials presented in [5], two different kinds of heuristics are performed concerning $mut(\vec{r})$:

H1: Restrict the selection of polynomials to those, which define a primitive polynomial $f(x) = 1 + x^{r_1} + x^{r_2}$ of degree r_2. If no such polynomial exists, select the next smaller existing. Here, $P_{mut}(\vec{r})$ gives the chance, that a given polynomial is exchanged by another one.

H2: Set $r_2 := l_1$ and select r_1 freely only considering the length of R_1, as mentioned above. Here, $P_{mut}(\vec{r})$ gives the chance that r_1 is modified. Note that $mut(\vec{q})$ changes r_2 indirectly by varying the length of R_1.

Heuristic H1 promotes the use of random sequences of high quality and therefore speeds up the convergence of the EA. In contradiction to H1, heuristic H2 does not force the exclusive use of m-sequences but considers a greater number of polynomials to be applied on the individuals. Here, the possibility of finding non-primitive polynomials yielding good results is given.

The flow of the algorithm is similar to the one described in Section 2.3.6.

Parameter Settings

In the EA a very small population of $N = 4$ individuals is used, i.e. $(4 + 4)$-method and $(4, 4)$-selection. This combination produces a high selection pressure and aims at a fast convergence. To prevent the "bottleneck effect", i.e. the occurance of most similar or even identical individuals within a population, a very high mutation rate concerning the weight vector ($P_{mut}(\vec{q}) = 1.0$) is used. In doing so, the creation of new weight combinations within the weight vectors is promoted. Of course, this also leads to an increased deterioration of some individuals in each iteration step. However, practice has shown that this effect does not affect the procedure in a negative way. The crucial point in this combination is that (N, μ)-selection will eliminate low-fitness individuals while the best individuals are protected through the $(N + \lambda)$-method.

At the beginning of an EA run the initial population is generated randomly. To get starting points that are not too bad the constant values 0 and 1 are excluded from being assigned to components of $\vec{q}$. The initial population is further improved by applying numerical hillclimbing [98] with respect to a subset $I_j \subset I$ of inputs to the weight vector of each of the μ fittest individuals, whereby

$$|I_1| = \ldots = |I_\mu| = \lceil \frac{n}{\mu} \rceil$$

and

$$I_1 \cup ... \cup I_\mu = I.$$

The resulting continuous values are rounded with respect to the chosen $(2, b)$-valued weight set afterwards.

For the first stage of the EA, it turns out that a number of $M_1 = 6400$ iterations is sufficient to obtain discrete weight vectors of high quality for all circuits considered. The most recent population of the first stage will form the initial population of the second stage of the EA. If the weight vectors started with are sufficiently good, the second stage may already be terminated after a number of $M_2 = 100$.

name	[98]			$W_{2,b}$			$W_{3,b}$			$W_{4,b}$		
	TL	*UF*	*WS*	*TL*	*UF*	*WS*	*TL*	*UF*	*WS*	*TL*	*UF*	*WS*
c2670	4703	0	-	9344	0	$W_{2,2}$	6624	0	$W_{3,2}$	4960	0	$W_{4,0}$
c7552	3422	0	-	12.6k	0	$W_{2,2}$	4992	0	$W_{3,1}$	3200	0	$W_{4,1}$
s9234.1	20096	3	-	36.1k	15	$W_{2,0}$	36.1k	2	$W_{3,0}$	36.1k	1	$W_{4,0}$
s13207.1	10.7k	0	-	14.7k	0	$W_{2,1}$	14.7k	0	$W_{3,1}$	13.4k	0	$W_{4,1}$
s15850.1	17.3k	0	-	32.6k	0	$W_{2,1}$	32.3k	0	$W_{3,0}$	18.5k	0	$W_{4,0}$
s38584.1	21.1k	0	-	36.1k	5	$W_{2,0}$	36.1k	1	$W_{3,0}$	36.1k	1	$W_{4,0}$

Table 6.15 Impact of rounding exact weights

Experimental Results

The implementation was done in *C++* using the OBDD package of [13]. For the experiments some of the ISCAS benchmark circuits [16, 15] are considered that contain a lot of random pattern resistant faults, so called "hard faults". In case of sequential circuits only their combinational parts are considered. These hard faults are identified heuristically by a fault simulation with equiprobable random pattern sequences of the lengths 1024 for the combinational circuits and 8196 the sequential circuits, respectively. Additionally, redundant faults are excluded from the fault list.

In the following tables *name* gives the name of the benchmark. *WV*, *WS*, *TL*, and *UF* denote the number of weight vectors the number of weight sets, the test length, and number of undetected faults, respectively.

Rounding of Exact Weights

Table 6.15 shows the results obtained after rounding the optimized exact input probabilities of [98] with respect to weight sets $W_{a,b}$, $a \in \{2,3,4\}$, $b \in \{0,1,2\}$ and the hardware of [155]. The second column gives the results obtained using exact weights [98]. From all weight sets of identical cardinality, those yielding the best result were considered. Obviously, the results become worse the fewer weights are used. This demonstrates trade-off between size of the testing hardware and quality of the resulting test.

The results concerning the weight set $W_{2,b}$ of Table 6.15 can be improved by using the first stage of the described EA. In Table 6.16 the results of these runs are given. Note, that TL includes the number of equiprobable random patterns. The results confirm the advantage of using EAs. For instance, in

name	EA, first stage		
	TL	*UF*	*WS*
c2670	5984	0	$W_{2,2}$
c7552	4736	0	$W_{2,2}$
s9234.1	36.1k	7	$W_{2,1}$
s13207.1	9184	0	$W_{2,0}$
s15850.1	17.8k	0	$W_{2,2}$
s38584.1	16.8k	0	$W_{2,0}$

Table 6.16 Improvement by the EA's first phase

name	EA			[88]			[130]			[82]		
	WV	*TL*	*UF*	*WV*	*TL*	*UF*	*WV*	*TL*	*UF*	*WV*	*TL*	*UF*
c2670	2	4960	0	5	3840	0	6	7552	0	2	16k	0
c7552	2	3200	0	7	6400	0	21	18.8k	0	2	256k	0
s9234.1	2	36.1k	1	-	-	-	9	23.1k	0	-	-	-
s13207.1	2	9184	0	-	-	-	3	7532	0	-	-	-
s15850.1	2	17.8k	0	-	-	-	5	14.5k	0	-	-	-
s38584.1	2	16.8k	0	14	18.9k	0	3	14.5k	0	-	-	-

Table 6.17 Comparison to other approaches

the case of circuit *s38584.1* the rounding of exact weights leads to a significant deterioration of the random testability. However, applying the EA to circuit *s38584.1* yields a result which is even better than the result obtained using exact weights.

In Table 6.17 the best EA results are compared with those obtained in [82, 88, 130]. The high quality of the results is evident, especially considering that the other approaches use a large number of weight vectors, e.g. for circuit *c7552*. Note that the approaches of [88, 130] can generally not be used in combination with a $(2, b)$-valued WRPG, as the latter restricts the range of applicable weight vectors to the equiprobable weight vector and exactly one additional $(2, b)$-valued weight vector, $b \geq 0$.

name	EA, second stage									
	H1					H2				
	WS	*UF*	*TLW*	*TLU*	*FE*	*WS*	*UF*	*TLW*	*TLU*	*FE*
c1908	$W_{2,1}$	0/5	1005	1379	98.72	$W_{2,1}$	0/0	967	5416	**100.00**
c2670	$W_{2,2}$	0/0	2261	897	**100.00**	$W_{2,2}$	0/0	2496	944	100.00
c3540	$W_{2,0}$	0/10	757	1642	99.70	$W_{2,0}$	0/0	1102	2728	**100.00**
c7552	$W_{2,2}$	9/0	10000	7110	99.88	$W_{2,2}$	0/1	3828	1140	**99.99**
s9234.1	$W_{2,1}$	0/3	28919	29436	**99.95**	$W_{2,0}$	7/0	32768	43176	99.89
s15850.1	$W_{2,2}$	0/1	13974	59715	99.99	$W_{2,2}$	0/0	12711	25556	**100.00**
s38584.1	$W_{2,0}$	0/13	3783	17387	99.96	$W_{2,0}$	0/2	8334	40660	**99.99**

Table 6.18 Determine the BIST structure by the EA's second phase

Evolution of $(2, b)$-valued WRPGs

The approach in [82] requires a considerably smaller hardware-overhead than all other hardware approaches described so far. So the goal is to verify the results optimized according to φ_r and applied to [155] with an $(2, b)$-valued WRPG. Thereby the search space of the EA increases considerably. Experiments show that it is necessary to have a good population as a starting point. The results of the EA's second step are given in Table 6.18. *WV* is equal to 2 for all numbers reported. TLW denotes the number of weighted random patterns, whereas TLU denotes the number of unweighted random patterns. The maximal number of unweighted pattern is set to five times the number of weighted pattern. The notation in column UF means "number of undetected hard faults / number of undetected non hard faults". As can be seen, for all circuits considered, a good result is received either with heuristic H1 or with heuristic H2: For each circuit all target faults, i.e. all hard faults, are detected. Considering all non redundant faults 100% or almost 100% are detected by the generated BIST structure.

A comparison of the quality results in Table 6.18 with those of others (see Table 6.17) can not be done directly, since there are more than two weight vectors used. Therefore these results do not match the minimal hardware-requirements used here. On the other hand, the EA easily outperforms the results of [82], given in Table 6.17.

Conclusions

The problem of optimizing weighted random pattern test in a BIST environment has been considered.

- Combining a genetic approach together with an OBDD based methods, it was possible to obtain a highly efficient BIST scheme with almost no area penalty.
- The EA consists of two stages:
    1. In a first stage of the EA high quality discrete input probabilities are computed based on *exact* fault detection probabilities.
    2. To efficiently realize the discrete input probabilities in hardware, a BIST scheme for the realization of (only) two weight sets is optimized in the second stage of the EA.
- The experiments demonstrated that patterns derived from this hardware result in fault efficiencies of (nearly) 100%. Comparison to the best previously known methods have been given.
- The runtime of the EA ranges from a few seconds for the small examples up to several CPU hours for the larger ones.

6.5 GUIDELINES FOR CAD APPLICATIONS

For EA based tools to make an impact in the CAD arena it is crucial that the developed algorithms are evaluated using the criteria commonly employed in the field. Specifically, the following guidelines should be followed:

1. Test examples should be *large*, reflecting the size of real-world problem instances. And whenever possible, benchmark data sets should be used to allow comparisons to other approaches.
2. Whenever possible, comparisons should be done to the existing state-of-the-art algorithms, considering both solution quality and runtime. For an EA based approach to be of practical interest, it is necessary and sufficient that it is competitive to state-of-the-art with respect to either solution

quality or runtime. While the partitioning algorithm and the SPG algorithm presented in Section 6.3 as well as the testing algorithms of Section 6.4 are competitive with respect to both criteria, the remaining algorithms presented in Section 6.3 and the algorithms of Section 6.2 are competitive on solution quality only.

3. Solution quality and runtime should be measured fairly, allowing comparisons to any other approach. I.e. runtime should be measured in terms of actual CPU time rather than e.g. number of points visited in the search space. And the total runtime should be measured, instead of e.g. the runtime until the occurrence of the last improvement.

4. Finally, it is important that a fixed set of parameters for the EA is used for all problem instances, i.e. problem specific tuning should not be performed[8]. If the designer needs to run the EA several times with different parameter settings in order to obtain a satisfactory performance, then this is a drawback of the EA, which should be taken into account when comparing runtime to that of another algorithm which was executed only once.

This set of guidelines may seem rather obvious to the reader, and is also complied to in the papers reviewed here. But unfortunately, it is too easy to find papers on EAs for VLSI CAD not complying with the above. While some of these papers may offer a theoretical contribution, they tend to do a disservice to the field by implying that the EA can not stand a fair comparison.

6.6 SUMMARY

In this chapter different EA applications for VLSI CAD have been presented. All in all, it has been demonstrated that many problems can be solved efficiently by EAs with respect to quality of the final result. The approaches based on genetic principles could in most cases improve on the results obtained by other heuristics.

On the other hand EA methods are not very effective when runtime is the main criterion. It turned out that in all applications the integration of problem

[8]In the case of an adaptive EA (adjusting parameter values dynamically), the initial parameter values, including values of parameters introduced to control the adaptive scheme, should be fixed.

specific knowledge (in the form of encoding or problem specific heuristics) is necessary to become competitive to other approaches.

Some general guidelines that should be followed when applying EAs in VLSI CAD have been summarized in the previous section.

7

HEURISTIC LEARNING

7.1 INTRODUCTION

As a result of the previous chapters we have seen that often the EA principles work well and get high quality results. Unfortunately, often the runtimes (especially for large problem instances) are overly long. For this in this chapter a model for heuristic learning is presented [37]. The idea of this approach is to create heuristics by an EA that obtain good results and in addition have good runtime behavior.

Due to the high complexity of the design process in VLSI CAD often heuristics are used. These heuristics are developed by the designer himself. But they also often fail for specific classes of circuits. Thus, it would help a lot, if the heuristics could learn from previous examples, e.g. from benchmark examples.

Notice that the EA presented in the following is not directly applied to the problem instance to be optimized (as the EAs discussed in the previous chapters). Instead the EA is used to develop a heuristic. The main difference to the approaches presented so far is that the time needed for the EA run (that might also be large for the EA in the following) is only invested once during the learning phase.

We first introduce the learning model and then describe an application of the model in more detail.

7.2 THE LEARNING MODEL

In the following it is assumed that the problem to be solved has the following property: There is defined a non empty set of optimization procedures that can be applied to a given (non-optimal) solution in order to further improve its quality. These procedures are called *Basic Optimization Modules* (BOMs).

These BOMs are the basic modules that will be used. Each heuristic is a sequence of BOMs. The goal of the approach is to determine a good (or even optimal) sequence of BOMs such that the overall results obtained by the heuristic are improved. The assumption concerning the problem is valid for all problems that can efficiently be optimized by greedy or hillclimbing heuristics. Notice that each "standard" heuristic already developed for the problem to be solved can also be integrated as a BOM, e.g. for finding a good starting point for the other operators.

As for standard EAs we here also assume a *finite set of finite strings* of fixed length over a given universe. This set is also called the *population*. For the considered problem that is to be solved by the EA an *objective function* is used that measures the *fitness* of each element. The main operations of EAs on strings are the standard operators introduced in Chapter 2:

Reproduction: Copying strings according to their fitness.

Crossover: Construction of a new element x from two parents y_1 and y_2, where the first part up to a cut position i is taken from y_1 and the second part is taken from y_2.

Mutation: Construction of a new element from a parent by copying the whole element and randomly changing its value at mutation position i.

Remark 7.1 Notice that here also more complex operators can be used, but they are left out for simplicity of the presentation.

Multi-valued strings are used, i.e. strings that may assume various values from a fixed finite range.

Remark 7.2 It does not influence the model, if the strings are considered over a two-valued alphabet. In this case an encoding must be used and method must be applied how to handle invalid solutions.

The set of BOMs defines the set H of all possible heuristics that are applicable to the problem to be solved in the given environment. H may include problem specific heuristics but can also include some random operators. To each BOM $h \in H$ a cost function $cost : H \times E \rightarrow \mathbf{R}$ is associated, where $\mathbf{R}$ denotes the real valued numbers and E denotes the set of examples. *cost* estimates the resources that are needed for a heuristic. (If we aim at fast heuristics a heuristic h with a large value $cost(h)$ should be avoided (if possible).) The fitness *fit* of a string $s = (h_1, h_2, \ldots, h_l)$ of length l (representing a heuristic composed from l BOMs) is measured by

$$fit(s) = c_c / fit_c(s) + c_q \cdot fit_q(s).$$

The cost fitness

$$fit_c(s) = \sum_{i=0}^{\#\ of\ examples} \sum_{j=0}^{l-1} cost(h_j, example_i)$$

of string s has to be minimized and the quality fitness

$$fit_q(s) = \sum_{i=0}^{\#\ of\ examples} quality(example_i)$$

of string s has to be maximized. c_c and c_q are problem specific constants.

The cost fitness measures the cost for the application of the string. If this cost is relatively high the resulting heuristic will take long time. If the heuristic has a low cost fitness it will terminate quickly.

The quality fitness measures the quality of the heuristic that is represented by the string s by summing up the results for a given *set of examples*. This set is also called the *training set*. Obviously the choice of the examples largely influences the quality of the resulting heuristic. Here the designer has to select a representative set of benchmarks. If this set cannot be determined, the EA can be run on a large set of "arbitrary" functions. The function *quality* measures the quality of the result with respect to the given problem. This function is typically the fitness function that is used in "normal" EAs.

The constants c_c and c_q are used to influence the primary goal of the heuristic: If c_c is set to 0 the EA will only optimize the heuristic with respect to the quality of the result, i.e. it will not care about the expenditure of the BOMs. If c_q is set to a small value the EA will determine a very fast heuristic, but the quality of the result will not be very good. Using these parameters the designer can influence the trade-off between runtime and quality and he can determine the primary goal of the EA: Should the heuristic focus on fast runtime or on good results?

7.3 A CASE STUDY: MINIMIZATION OF DECISION DIAGRAMS

In this section one problem is studied in more detail: decision diagrams have been used in Chapter 6 for evaluating the fitness function in FPRM minimization. In the following methods are studied to minimize the diagram itself.

7.3.1 Decision Diagrams

Decision Diagrams (DDs) are often used in CAD systems for efficient representation and manipulation of Boolean functions [110, 19, 151]. (In Section 6.2 they have also been proposed as a data structure for logic synthesis.) The most popular data structure is the *Ordered Binary Decision Diagram* (OBDD) [18]. OKFDDs are a generalization of OBDDs and OFDDs as well and try to combine the advantages of both representations by allowing the use of Shannon decompositions and (positive and negative) Davio decompositions in one and the same DD [36].

As well-known for OBDDs [18] and OFDDs [8] OKFDDs are very sensitive to the variable ordering.

Example 7.1 Let $f = x_1x_2 + \ldots + x_{2n-1}x_{2n}$. If the variable ordering is given by $(x_1, x_2, \ldots, x_{2n})$ the size of the resulting OBDD is $2n$. On the other hand if the variable ordering is chosen as $(x_1, x_{n+1}, x_2, x_{n+2}, \ldots, x_{2n})$ the size of the OBDD is $\Theta(2^{n-1})$. Thus the number of nodes in the graph varies from linear to exponential depending on the variable ordering. In Figure 7.1 the OBDDs of the function $f = x_1x_2 + x_3x_4 + x_5x_6$ with variable orderings $x_1x_2x_3x_4x_5x_6$ and $x_1x_3x_5x_2x_4x_6$ are illustrated. The left (right) outgoing edge of each node x_i denotes the cofactor to 1 (0). As can be seen the choice of the variable ordering largely influences the size of the OBDDs.

Analogously to OBDDs the computation of an optimal variable ordering for OKFDDs is NP-hard [12] and the best known algorithm has runtime $O(n^2 \cdot 3^n)$

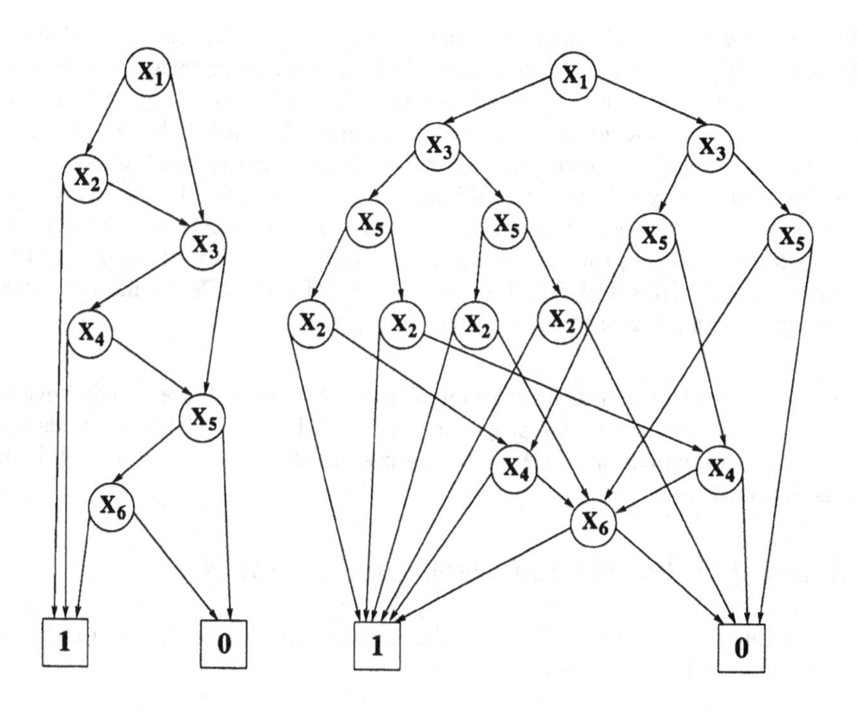

Figure 7.1 OBDDs for function $f = x_1x_2 + x_3x_4 + x_5x_6$

[66, 86], where n denotes the number of variables. In addition to the position of a variable in the ordering a so-called decomposition type has to be chosen for OKFDDs. Thus, there is a need for heuristics to choose a suitable variable ordering and decomposition type list for OKFDDs.

In the last few years many authors presented heuristics for finding good variable orderings for OBDDs [110, 68, 114, 67, 69]. The most promising methods are based on *dynamic variable ordering* [70, 132, 122]: OBDDs for some Boolean functions could be constructed for which all other (topology oriented methods) failed. In [36] it has been shown that dynamic variable ordering methods for OBDDs can also be applied to OKFDDs.

Recently, a new method based on evolutionary algorithms has been proposed for OKFDD minimization [44]. The major drawback of this approach is that in general it obtains good results with respect to quality of the solution, but the runtimes are often much larger than that of classical heuristics.

In the following an EA based approach to learn heuristics for OKFDD minimization is presented. The EA learns heuristics starting from some simple basic operations that are (mainly) based on dynamic reordering. The learning environment is a set of benchmark examples, also called the *training set*. By experiments it is shown that the EA designs heuristics that improve the results obtained by iterated DTL-sifting [36] by about 10%. Furthermore, the runtimes of the developed heuristics are low, since the costs of the heuristic are minimized during the learning process. Notice once more that due to the reason that OBDDs and OFDDs are a subset of OKFDDs all methods can directly be applied also to these restricted types of DDs.

Before we start with the description of the EA approach we briefly review the essential definitions and properties of OKFDDs [36]. Then methods for heuristical minimization of OKFDDs are presented that are used as BOMs in this approach.

Kronecker Functional Decision Diagrams

First a representation form that is a slight extension of OFDDs (as introduced in Section 6.2.1) are defined:

A *Kronecker Functional Decision Diagram* (KFDD) is a graph-based representation of a Boolean function $f : \mathbf{B}^n \rightarrow \mathbf{B}$ over the variable set X_n (similar to BDDs), where one of the following three decompositions is carried out in each node:

$$f = \overline{x}_i f_i^0 + x_i f_i^1 \qquad \textit{Shannon } (S) \qquad (7.1)$$

$$f = f_i^0 \oplus x_i f_i^2 \qquad \textit{positive Davio } (pD) \qquad (7.2)$$

$$f = f_i^1 \oplus \overline{x}_i f_i^2 \qquad \textit{negative Davio } (nD) \qquad (7.3)$$

(f_i^0 defined by $f_i^0(x) := f(x_1, .., x_{i-1}, 0, x_{i+1}, .., x_n)$ for $x = (x_1, x_2, \ldots, x_n) \in \mathbf{B}^n$ denotes the *cofactor* of f with respect to $x_i = 0$, f_i^1 denotes the cofactor for $x_i = 1$ and f_i^2 is defined as $f_i^2 := f_i^0 \oplus f_i^1$, $\oplus$ being the Exclusive OR operation.) If the underlying graph is *ordered* the corresponding KFDD is called an OKFDD.

Decomposition types are associated to the variables in X_n with the help of a *Decomposition Type List* (DTL) $d := (d_1, \ldots, d_n)$ where $d_i \in \{S, pD, nD\}$, i.e. for each variable one fixed decomposition is chosen.

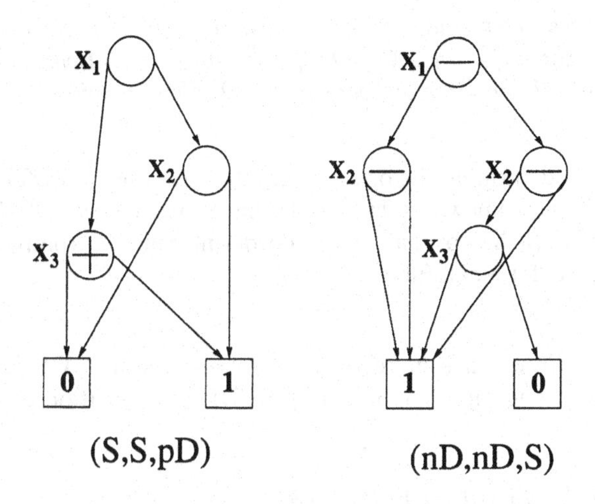

Figure 7.2 OKFDDs for function $f = x_1x_2 + \bar{x}_1x_3$

On OKFDDs reductions can be defined. From [36] it is known that reductions can be used to define canonical representations for not only OBDDs and OFDDs, but also for OKFDDs.

The size of OKFDDs is very sensitive to the variable ordering and the choice of the *Decomposition Type List* (DTL).

Example 7.2 Let $f = x_1x_2 + \bar{x}_1x_3$. In Figure 7.2 two OKFDDs representing f are given. Both OKFDDs have the same variable ordering, but the DTL differs, i.e. the OKFDD on the left hand side has DTL (S, S, pD), while the OKFDD on the right hand side has DTL (nD, nD, S). In the figure an empty node denotes a Shannon decomposition, + and – denote positive and negative Davio decomposition. The size of the resulting graphs are 3 and 4, respectively.

Remark 7.3 Even though in the example the influence of the choice of the DTL is small for other (more difficult examples) the size may vary from linear to exponential [7].

Now the following problem is considered that will be solved using the EA:

How can we determine a good heuristic to perform variable ordering and DTL choice for an OKFDD representing a given Boolean function f such that the number of nodes in the OKFDD is minimized?

The problem to determine the optimal variable ordering and DTL seems to be very hard. The best known exact solution has runtime $O(n^2 \cdot 3^n \cdot 3^n)$, since for each possible DTL the optimal variable ordering must be computed and thus it is only feasible for small functions.

Remark 7.4 Notice once more that we do *not* optimize OKFDDs by EAs. Instead the heuristic that is applied to OKFDD minimization is optimized.

Dynamic Variable Ordering

It is well-known for decision diagrams that the sizes can be minimized by exchanging adjacent variables [70]. The exchange is performed very quickly since only edges must be redirected[1]. This exchange is the basic operation for different algorithms for dynamic variable ordering, e.g the sifting algorithm, DTL-sifting, window permutation, and exact minimization algorithms [132, 86, 36]. In the following the algorithms that are used as BOMs in the approach in the next section are briefly described:

Sifting (S) [132]: By the sifting algorithm, the variables are sorted into decreasing order based on the number of nodes at each level and then each variable is moved through the OKFDD in order to locate its local optimal position while all other variables and the DTL remain fixed.

Siftlight (L): Siftlight is a restricted form of sifting that does not allow the algorithm to do any hillclimbing, i.e. the variables are directly located in the next minimum. (The algorithm is much faster than "normal" sifting, but in general the results are worse.)

DTL-sifting (D) [36]: By DTL-sifting the variables are sorted into decreasing order based on the number of nodes at each level. Then each variable traverses for all three different decomposition types the OKFDD analogously to *sifting*. Thus, the variable ordering and DTL are dynamically optimized.

[1] In this approach complemented edges are also used.

Exact (E): Perform the exact minimization algorithm for OKFDDs for only three adjacent levels (=window). A window is chosen for optimization, if the sum over all nodes in these levels is maximal.

Inversion (I): The variable ordering of the OKFDD is inverted.

7.3.2 Evolutionary Algorithm

In this section the EA is described that is applied to the problem given above.

Representation

In the application a multi-valued encoding is used, for which the problem can easily be formulated. Each position in a string represents an application of a BOM. Thus, a string represents a sequence of heuristics. If a string has n components at most n applications of BOMs are possible. (This upper bound is set by the designer and limits the runtime of the heuristic.) Thus, each element of the population corresponds to an n-dimensional multi-valued vector. Using this multi-valued encoding each string represents a valid solution.

In the following six-valued vectors over the alphabet $\{S, L, D, E, I, N\}$ are considered:

- S, L, D, E and I represents sifting, siftlight, DTL-sifting, exact and inversion, respectively.
- N (no operation) means that no operation is performed. The operation N takes no resources and thus the costs of the resulting heuristic can be minimized.

For simplicity we restrict to these alternatives, since they have shown to work very well in this application.

Objective Function and Selection

As an *objective function* that measures the *fitness* of each element the heuristics are applied to benchmark *training sets*. Obviously, the choice of the benchmarks largely influences the results. On the other hand the designer can create several different heuristics for different types of circuits, e.g. a fast but simple heuristic for very large problem instances or a relative "time consuming" heuristic for small examples. Furthermore, tuning of heuristics towards specific properties of the functions, like symmetry, is possible. The fitness of each element is computed analogously to the formula given above:

- For the calculation of function *quality* the OKFDD is constructed using the initial variable ordering and Shannon decomposition in each node.

  Then the heuristic represented by the considered element is performed and the number of nodes is counted. The function is then given by the equation $quality = 1/nodes$, since the number of nodes has to be minimized.

- Function *cost* measures the computation time that is used to evaluate an element, i.e. the time that is needed to reorder the OKFDD, in 10^{-2} CPU seconds.

- The choice of the parameters c_c and c_q is discussed below in more detail.

Again, selection is performed by *roulette wheel selection* combined with *steady-state-reproduction.*

Genetic Operators

As genetic operators standard *reproduction*, *crossover* and *mutation* and some slightly modified operators are used. All operators are directly applied to six-valued strings of finite length that represent elements in the population. The parent(s) for each operation is (are) determined by the mechanisms described above. All genetic operators only generate valid solutions, if they are applied to multi-valued strings.

Algorithm

Using the genetic operators the algorithm works as follows:

1. The initial population *pop* of size 10 is generated randomly and the length of the strings is set to 20. (Notice, the length of the strings is only an upper bound for the number of BOMs that are applied to an OKFDD since the "empty" BOM N gives no costs.)

2. Then $\frac{pop}{2}$ elements are generated by the genetic operators that are applied with a corresponding probability. The newly created elements are then

```
evolutionary_algorithm (training set):
    generate_random_population_of_heuristics ;
    calculate_fitness_by_evaluation_on_training_set ;
    do
        apply_operators_with_corresponding_probabilities ;
        calculate_fitness_by_evaluation_on_training_set ;
        update_population ;
    while (not terminal case) ;
    return best_heuristics ;
```

Figure 7.3 Sketch of learning algorithm

mutated with a probability of 15%. After each iteration the size of the population is constant.

3. If no improvement is obtained for 50 generations the algorithm stops.

A sketch of the algorithm is given in Figure 7.3. As can easily be seen the overall flow of the algorithm is very similar to a standard GA (see Chapter 2). The important difference is that the parameter to the function is not a single problem instance to be optimized; instead a training set is used.

7.3.3 Experimental Results

In this section results of experiments are presented that were carried out on a *SUN Sparc 20* workstation. All runtimes *time* are given in CPU seconds. The benchmark functions are taken from LGSynth91 [160]. The best results are given in bold in the following.

In a first series of experiments a heuristic for OKFDD minimization on a small *training set* that is composed of only five functions has been developed. The learning time takes about 12-14 CPU hours. Notice that this time for the learning task has to be spent only once. The resulting heuristics are very fast.

The results for size and runtime if the created heuristic is applied to the training set elements are given in Table 7.1 and 7.2, respectively. *in* (*out*) denotes the number of inputs (outputs) of the corresponding benchmark and *size* gives the number of nodes. DTL denotes the results after DTL-sifting iteratively applied

name	*in*	*out*	DTL	EA1	EA2	EA3	EA4
add6	12	7	24	68	25	**23**	**23**
alu2	10	6	138	157	127	120	**119**
frg1	28	3	**70**	79	75	74	72
x6dn	39	5	217	244	203	**196**	**196**
Z5xp1	7	10	**28**	41	**28**	**28**	**28**

Table 7.1 Representation sizes for training set

name	*in*	*out*	DTL	EA1	EA2	EA3	EA4
add6	12	7	0.3	< 0.1	0.2	0.3	0.6
alu2	10	6	1.0	0.1	0.9	1.2	2.4
frg1	28	3	1.2	< 0.1	0.9	1.3	1.6
x6dn	39	5	1.7	0.1	2.2	2.7	4.4
Z5xp1	7	10	0.2	0.2	0.4	0.4	1.1

Table 7.2 Runtimes for training set

until no further improvement could be obtained. (Thus, DTL-sifting implicitly makes use of a more powerful "do-until"-operator. Nevertheless the results demonstrate that the method obtains better results with "weaker" operators.) The four rightmost columns show the results after applying the newly developed heuristics that are learned by the EA for varying parameter settings of c_c. For each EA c_q is fixed by a problem specific constant factor and c_c is chosen in the range from 0 to 1; the exact settings of c_c are $c_c = 1$, $c_c = \frac{1}{10}$, $c_c = \frac{1}{100}$, and $c_c = 0$ for EA1, EA2, EA3, and EA4, respectively. The larger c_c is chosen the larger is the influence of the timing aspect.

The results obtained by EA1, where runtime is the main optimization goal, in all cases are worse than DTL-sifting, but the corresponding runtimes are much better, i.e. never larger than 0.2 CPU seconds. If the influence of the timing aspect is decreased, the quality of the results gets higher (in most cases equal or better than DTL-sifting). EA2 produces better results than DTL-sifting on average and the runtimes are also often faster. (In the cases where EA2 is slower it is able to improve the results with respect to quality.) The heuristic that is developed by EA4 does not consider any timing optimization. As can

name	*in*	*out*	DTL	EA1	EA2	EA3	EA4
addm4	9	8	126	163	130	126	**125**
apex2	39	2	310	553	274	**243**	273
apex7	49	37	264	300	225	230	**220**
bc0	26	11	431	522	426	**423**	**423**
bcd	26	38	565	574	568	569	**561**
cm85a	11	3	35	35	26	**20**	**20**
chkn	29	7	260	324	246	**228**	246
cps	24	109	573	726	**545**	563	548
ex5	8	63	**234**	241	304	**234**	281
ex7	16	5	75	77	62	**60**	**60**
gary	15	11	286	297	282	**280**	292
in7	26	10	**64**	83	73	76	75
m181	15	9	**53**	54	55	**53**	56
mlp4	8	8	**107**	134	108	108	108
pdc	16	40	**572**	605	649	639	663
rd73	7	3	30	30	**21**	**21**	**21**
risc	9	8	**56**	65	**56**	**56**	**56**
sqn	7	3	**47**	55	51	51	51
t1	21	23	**103**	114	126	119	120
tial	8	31	550	613	489	473	**454**
ts10	22	16	**131**	145	160	160	**131**
vg2	25	8	186	193	95	94	**76**

Table 7.3 Representation sizes for application to new benchmarks

be seen EA4 determines the best OKFDD sizes on average on the training set. The runtimes are still in an acceptable range, i.e. it takes only a few seconds.

In a next series of experiments the developed heuristics are applied to new benchmarks that were not included in the training set, i.e. functions that were unknown during the optimization process of the EA. The results are given in Table 7.3 and 7.4, respectively. As can easily be seen the timing behaviour of the heuristics is similar to the first experiment. The application of DTL-sifting obtains best solutions for about 40% of the considered Boolean functions. For EA3 and EA4 the best solutions are obtained for nearly 50%, even though the resulting heuristics were not trained on these functions. Even EA2 improves DTL-sifting with respect to quality, and the runtimes of EA2 are nearly the same as DTL-sifting. EA1 often fails, but the execution time is (much) smaller

name	*in*	*out*	DTL	EA1	EA2	EA3	EA4
addm4	9	8	0.6	0.1	0.8	1.0	2.1
apex2	39	2	147.8	22.6	153.0	198.1	178.3
apex7	49	37	2.3	0.3	2.7	3.1	4.8
bc0	26	11	2.3	0.3	2.5	3.1	4.5
bcd	26	38	6.1	0.8	6.2	8.5	11.2
cm85a	11	3	< 0.1	< 0.1	0.1	0.1	0.2
chkn	29	7	16.8	1.8	11.7	15.3	17.2
cps	24	109	10.7	1.1	7.0	9.3	11.2
ex5	8	63	0.6	0.1	0.9	1.1	2.9
ex7	16	5	0.8	0.1	1.0	1.3	1.8
gary	15	11	0.9	0.1	1.1	1.5	2.3
in7	26	10	0.9	0.2	1.3	1.7	1.9
m181	15	9	0.6	0.1	0.8	1.0	1.9
mlp4	8	8	0.5	0.2	1.0	1.3	2.7
pdc	16	40	1.6	0.8	5.8	7.9	10.2
rd73	7	3	0.1	< 0.1	0.1	0.2	0.6
risc	9	8	0.1	< 0.1	0.1	0.2	0.6
sqn	7	3	0.1	< 0.1	0.1	0.2	0.4
t1	21	23	0.1	0.1	0.4	0.6	0.7
tial	8	31	6.4	0.8	4.8	9.7	10.0
ts10	22	16	2.0	2.4	7.2	10.3	10.9
vg2	25	8	0.5	0.1	0.9	1.2	1.8

Table 7.4 Runtimes for application to new benchmarks

than that of all other strategies. It is important to notice that the learned heuristic EA4 is never more than 20% worse than DTL-sifting. On the other hand DTL-sifting gets stuck very early in some cases and gets results that are more than 50% worse (see e.g. *vg2*).

The training set in the first run only consists of five benchmarks. The evaluation of the resulting heuristic on benchmarks that were unknown during the training phase showed that the EA "failed" for some examples. In a next experiment the training set is extended by three further examples that obtained unsatisfying results, i.e. functions *pdc, t1* and *sqn* from Table 7.3. For these functions DTL-sifting created smaller OKFDDs. The goal is to improve the heuristics by adapting them to the *extended* training set. The results of this experiment

name	*in*	*out*	DTL	EA1	EA2	EA3	EA4
add6	12	7	24	41	24	**23**	**23**
alu2	10	6	138	149	137	**122**	124
frgl	28	3	**70**	94	77	75	72
x6dn	39	5	217	232	215	213	**203**
Z5xp1	7	10	**28**	29	**28**	**28**	**28**
pdc	16	40	572	574	**564**	**564**	565
sqn	7	3	**47**	56	**47**	**47**	**47**
t1	21	23	103	111	110	103	**102**

Table 7.5 Representations sizes for extended training set

name	*in*	*out*	DTL	EA1	EA2	EA3	EA4
add6	12	7	0.3	< 0.1	0.2	0.2	0.6
alu2	10	6	1.0	0.1	0.5	1.1	2.2
frgl	28	3	1.2	< 0.1	1.1	1.1	2.4
x6dn	39	5	1.7	0.1	1.8	3.2	5.7
Z5xp1	7	10	0.2	0.1	0.2	0.3	1.2
pdc	16	40	1.6	0.2	4.5	6.2	12.9
sqn	7	3	0.2	< 0.1	0.1	0.1	0.5
t1	21	23	0.1	< 0.1	0.2	0.5	1.1

Table 7.6 Runtimes for extended training set

are given in Table 7.5 and 7.6, respectively. As can easily be seen the results (especially on the new examples) become much better than for DTL-sifting. All EAs now produce a very good number of nodes for the "worst case examples".

Finally, the new heuristics that were learned on the extended training set are applied to the remaining benchmarks (that were again unknown during the optimization). The results are given for EA3 and EA4 in Table 7.7. (EA3 and EA4 are chosen because of their performance concerning the quality of the results.) The experiments show that EA4 is now only in one case, i.e. *chkn*, worse than DTL-sifting (and the difference in this case is only one node). For **all**

name	*DTL*		*EA3*		*EA4*	
	size	*time*	*size*	*time*	*size*	*time*
addm4	126	0.6	126	0.9	**125**	2.4
apex2	310	148.4	323	170.9	**273**	270.1
apex7	264	2.3	259	2.9	**187**	7.4
bc0	**431**	2.3	**431**	2.3	**431**	7.3
bcd	565	6.1	**561**	7.1	**561**	13.4
cm85a	35	< 0.1	35	0.1	**33**	0.3
chkn	**260**	16.8	315	13.6	261	24.7
cps	573	10.7	**556**	7.9	560	18.9
ex5	**234**	0.6	**234**	0.9	**234**	2.6
ex7	**75**	0.8	**75**	1.1	**75**	2.2
gary	286	0.9	292	1.3	**279**	3.3
in7	**64**	0.9	78	1.5	**64**	3.0
m181	**53**	0.6	**53**	1.0	**53**	2.5
mlp4	**107**	0.5	108	1.1	**107**	3.1
rd73	30	0.1	30	0.2	**21**	0.7
risc	**56**	0.1	**56**	0.2	**56**	0.6
tial	550	6.4	533	5.7	**471**	13.4
ts10	**131**	2.0	**131**	8.1	**131**	13.2
vg2	186	0.5	187	1.1	**184**	3.6

Table 7.7 Application to new benchmarks II

other benchmarks EA4 is (much) better or equal to DTL-sifting. The runtimes are still reasonable, i.e. less than 20 CPU seconds for most examples.

All in all, the experiments have shown that the EA is able to design a heuristic that is more powerful than a "hand-designed" heuristic. The larger the training set is chosen the higher is the average gain in the quality of the results. This also enables the option to add new examples to the EA run and in this way to dynamically adapt the heuristic to new problem instances.

7.4 SUMMARY

In this chapter a model for learning heuristics based on EAs has been presented. The EA depends on a parameter set that influences the quality and the runtimes of the resulting heuristic. Thus, it is possible to design differing heuristics, e.g. a fast heuristic that obtains reasonable results or a high quality heuristic that may take more execution time.

The model has been applied to minimization of a specific type of decision diagram, called OKFDD. The EA learns the heuristic on a *training set* composed of examples of Boolean functions. The resulting heuristic works very well with respect to quality and costs on these examples. The application to new benchmarks that were unknown during the learning process demonstrates the efficiency of the new heuristic. Furthermore, the integration of new examples into the learning set improves the quality of the newly developed heuristic. This technique enables the designer to create problem specific heuristics, e.g. for a specific class of circuits.

It should be mentioned that the methods can be directly applied to more restricted classes of DDs, like OBDDs and OFDDs. Also the use of more powerful sifting operators (see e.g. [123, 115, 122]) can easily be incorporated. Further application of the model in areas of logic synthesis and DD minimization [37, 47, 75] obtained very promising results. In all cases the best previously known heuristic could be improved with respect to quality and runtime. It was especially interesting to observe that the newly created heuristic also performed very well on examples that were unknown during the learning process.

Also the model itself can easily be extended by allowing more powerful operators. The model as defined above does not allow the integration of loops or decisions. For harder problems it might become necessary to integrate these more complex approaches.

8

CONCLUSIONS

Three main conclusions are drawn from this work:

- First of all, the EA has a very large potential within VLSI CAD. The problems encountered in this field are extremely complex, which is exactly the situation in which the performance of the EA compares best to that of other methods. For some of the VLSI CAD problems considered in this book the presented EA approaches define the state-of-the-art.

- To make an impact in the CAD field, it is crucial that certain practical guidelines for performance evaluation are followed. It will always be difficult, and often impossible, to provide an absolute fair comparison of two algorithms with very different characteristics. However, applying the guidelines from Section 6.5 is likely to eliminate at least some of the problems often occurring, and will in any case improve the reputation of the EA in the CAD world.

- The main characteristic of the algorithms reviewed in this book is that they apply problem-specific knowledge in one way or another. Combinations with other heuristics, problem-specific representations and/or operators are the rule rather than the exception. This strongly suggests that incorporation of problem-specific knowledge is necessary for the EA to be competitive to the best existing approaches. Therefore, the traditional "pure" EA will most often have to be abandoned when competitive performance is the main objective.

In addition to these general issues, the basic components of an EA implementation has been presented. To obtain competitive performance, it is of paramount importance to incorporate existing problem-specific knowledge and/or heuristics in the EA environment. Since the EA naturally supports integration of problem-specific heuristics, it effectively constitutes an optimization environment enabling software reuse. support software reuse. Several examples have been given in Chapter 6 to demonstrate the importance of these points.

Finally, as an alternative to the classical EA applications a method to learn heuristics for VLSI CAD using EAs has been presented. The method is very general, since it applies to all problems for which greedy and hillclimbing heuristics can be applied.

EAs are very well suited for VLSI CAD problems, but they only show their full potential when problem-specific knowledge is appropriately utilized.

REFERENCES

[1] M. Abramovici and M.A. Breuer. On redundancy and fault detection in sequential circuits. *IEEE Trans. on Comp.*, 28(11):864–865, 1979.

[2] V.D. Agrawal, K.-T. Cheng, and P. Agrawal. CONTEST: a concurrent test generator for sequential circuits. In *Design Automation Conf.*, pages 84–89, 1988.

[3] V.D. Agrawal, K.-T. Cheng, D.D. Johnson, and T. Lin. Designing circuits with partial scan. *IEEE Design & Test of Comp.*, 5(4):8–15, 1988.

[4] T. Bäck. *Evolutionary Algorithms in Theory and Practice.* Oxford University Press, 1996.

[5] P.H. Bardell, W.H. McAnney, and J. Savir. *Built-In Test for VLSI: Pseudorandom Techniques.* John Wiley & Sons, 1987.

[6] B. Becker and R. Drechsler. OFDD based minimization of fixed polarity Reed-Muller expressions using hybrid genetic algorithms. In *Int'l Conf. on Comp. Design*, pages 106–110, 1994.

[7] B. Becker, R. Drechsler, and M. Theobald. On the expressive power of okfdds. *Formal Methods in System Design: An International Journal*, 11(1):5–21, 1997.

[8] B. Becker, R. Drechsler, and R. Werchner. On the relation between BDDs and FDDs. *Information and Computation*, 123(2):185–197, 1995.

[9] M. Blum, A.K. Chandra, and M.N. Wegman. Equivalence of free Boolean graphs can be decided probabilistically in polynomial time. *Information Processing Letters*, 10:80–82, 1980.

[10] K.D. Boese, A.B. Kahng, and G. Robins. High-performance routing trees with identified critical sinks. In *Design Automation Conf.*, pages 182–187, June 1996.

[11] B. Bollig, M. Löbbing, and I. Wegener. Simulated annealing to improve variable orderings for OBDDs. In *Int'l Workshop on Logic Synth.*, pages 5b:5.1–5.10, 1995.

[12] B. Bollig and I. Wegener. Improving the variable ordering of OBDDs is NP-complete. *IEEE Trans. on Comp.*, 45(9):993–1002, Sept. 1996.

[13] K.S. Brace, R.L. Rudell, and R.E. Bryant. Efficient implementation of a BDD package. In *Design Automation Conf.*, pages 40–45, 1990.

[14] M.A. Breuer and A.D. Friedman. *Diagnosis & reliable design of digital systems.* Computer Science Press, 1976.

[15] F. Brglez, D. Bryan, and K. Kozminski. Combinational profiles of sequential benchmark circuits. In *Int'l Symp. Circ. and Systems*, pages 1929–1934, 1989.

[16] F. Brglez and H. Fujiwara. A neutral netlist of 10 combinational circuits and a target translator in fortran. In *Int'l Symp. Circ. and Systems, Special Sess. on ATPG and Fault Simulation*, pages 663–698, 1985.

[17] S.D. Brown, R.J. Francis, J. Rose, and Z.G. Vranesic. *Field-Programmable Gate Arrays.* Kluwer Academic Publisher, 1992.

[18] R.E. Bryant. Graph - based algorithms for Boolean function manipulation. *IEEE Trans. on Comp.*, 35(8):677–691, 1986.

[19] R.E. Bryant. Symbolic Boolean manipulation with ordered binary decision diagrams. *ACM, Comp. Surveys*, 24:293–318, 1992.

[20] T.N. Bui and B.R. Moon. A fast stable hybrid genetic algorithm for the ratio-cut partitioning problem on hypergraphs. In *Design Automation Conf.*, pages 664–669, 1994.

[21] K.-T. Cheng and V.D. Agrawal. State assignment for initializable synthesis. In *Int'l Conf. on CAD*, pages 212–215, 1989.

[22] H. Cho, G.D. Hatchel, and F. Somenzi. Redundancy identification/removal and test generation for sequential circuits using implicit state enumeration. *IEEE Trans. on CAD*, 12(7):935–945, 1993.

[23] H. Cho, S. Jeong, F. Somenzi, and C. Pixley. Synchronizing sequences and symbolic traversal techniques in test generation. *Jour. of Electronic Testing: Theory and Applications*, 4:19–31, 1993.

[24] J.P. Cohoon, S.U. Hedge, W.N. Martin, and D. Richards. Distributed genetic algorithms for the floorplan design problem. *IEEE Trans. on CAD*, 10:484–492, 1991.

[25] J. Cong and Y. Ding. On area/depth trade-off in LUT-based FPGA technology mapping. *IEEE Trans. on VLSI Systems*, 2(2):137–148, 1994.

[26] T.H. Cormen, C.E. Leierson, and R.C. Rivest. *Introduction to Algorithms*. MIT Press, McGraw-Hill Book Company, 1990.

[27] F. Corno, P. Prinetto, M. Rebaudengo, and M.S. Reorda. Comparing topological, symbolic and GA-based ATPGs: an experimental approach. In *Int'l Test Conf.*, pages 39–47, 1996.

[28] F. Corno, P. Prinetto, M. Rebaudengo, and M.S. Reorda. GATTO: A genetic algorithm for automatic test pattern generation for large synchronous sequential circuits. *IEEE Trans. on CAD*, 15(8):991–1000, 1996.

[29] F. Corno, P. Prinetto, M. Rebaudengo, and M.S. Reorda. Partial scan flip flop selection for simulation-based ATPGs. In *Int'l Test Conf.*, pages 558–564, 1996.

[30] F. Corno, P. Prinetto, M. Rebaudengo, M.S. Reorda, and R. Mosca. Advanced techniques for ga-based sequential atpg. In *European Design & Test Conf.*, pages 375–379, 1996.

[31] O. Coudert. Two-level logic minimization: an overview. *Integration the VLSI Jour.*, (17):97–140, 1994.

[32] O. Coudert, H. Fraisse, and J.C. Madre. A breakthrough in two-level logic minimization. In *Int'l Workshop on Logic Synth.*, page P2b, 1993.

[33] P. Dasgupta, P. Mitra, P.P. Chakrabarti, and S.C. DeSarkar. Multiobjective search in vlsi design. In *VLSI Design Conf.*, pages 395–400, 1994.

[34] L. Davis. *Handbook of Genetic Algorithms*. van Nostrand Reinhold, New York, 1991.

[35] K. Dill and M.A. Perkowski. Minimization of GRM forms with a genetic algorithm. In *Genetic Programming Conference*, 1997.

[36] R. Drechsler and B. Becker. Dynamic minimization of OKFDDs. In *Int'l Conf. on Comp. Design*, pages 602–607, 1995.

[37] R. Drechsler and B. Becker. Learning heuristics by genetic algorithms. In *ASP Design Automation Conf.*, pages 349–352, 1995.

[38] R. Drechsler and B. Becker. Relation between OFDDs and FPRMs. *Electronic Letters*, 32:1975–1976, 1996.

[39] R. Drechsler and B. Becker. Overview of decision diagrams. *IEE Procedings*, 144:187–193, 1997.

[40] R. Drechsler, B. Becker, and N. Göckel. A genetic algorithm for minimization of fixed polarity Reed-Muller expressions. In *Int'l Conf. on Artificial Neural Networks and Genetic Algorithms*, pages 392–395, 1995.

[41] R. Drechsler, B. Becker, and N. Göckel. A genetic algorithm for variable ordering of OBDDs. In *Int'l Workshop on Logic Synth.*, pages P5c:5.55–5.64, 1995.

[42] R. Drechsler, B. Becker, and N. Göckel. A genetic algorithm for the construction of small and highly testable OKFDD circuits. In *Genetic Programming Conference*, pages 473–478, 1996.

[43] R. Drechsler, B. Becker, and N. Göckel. A genetic algorithm for variable ordering of OBDDs. *IEE Procedings*, 143(6):364–368, 1996.

[44] R. Drechsler, B. Becker, and N. Göckel. Minimization of OKFDDs by genetic algorithms. In *Int'l Symposium on Soft Computing*, pages B:271–B:277, 1996.

[45] R. Drechsler, B. Becker, and N. Göckel. A genetic algorithm for RKRO minimization. *Expert Systems with Applications: An International Journal*, 12(1):127–139, 1997.

[46] R. Drechsler and N. Göckel. Minimization of BDDs by evolutionary algorithms. In *Int'l Workshop on Logic Synth.*, 1997.

[47] R. Drechsler, N. Göckel, and B. Becker. Learning heuristics for OBDD minimization by evolutionary algorithms. In *Parallel Problem Solving from Nature, LNCS 1141*, pages 730–739, 1996.

[48] R. Drechsler, H. Hengster, H. Schäfer, J. Hartmann, and B. Becker. Testability of 2-level AND/EXOR expressions. In *European Design & Test Conf.*, pages 548–553, 1997.

[49] R. Drechsler, M. Theobald, and B. Becker. Fast OFDD based minimization of fixed polarity Reed-Muller expresssions. *IEEE Trans. on Comp.*, 45:1294–1299, 1996.

[50] S.E. Dreyfuss. An appraisal of some shortest-path algorithms. *Journal of the Operational Research Society*, 17:395–412, 1969.

[51] R.D. Eldred. Test routines based on symbolic logical statements. *Journal of the ACM*, 6(1):33–36, 1959.

[52] H. Esbensen. A macro-cell global router based on two genetic algorithms. In *European Design Automation Conf.*, pages 428–433, 1994.

[53] H. Esbensen. Computing near-optimal solutions to the steiner problem in a graph using a genetic algorithm. *NETWORKS*, 26:173–185, 1995.

[54] H. Esbensen. Finding (near-)optimal steiner trees in large graphs. In *Int'l Conference on Genetic Algorithms*, pages 485–491, 1995.

[55] H. Esbensen. Defining solution set quality. Technical report, UCB/ERL M96/1, University of Berkeley, 1996.

[56] H. Esbensen and E. Kuh. Design space exploration using the genetic algorithm. In *Int'l Symp. Circ. and Systems*, pages IV:500–IV:503, 1996.

[57] H. Esbensen and E. Kuh. A performance-driven IC/MCM placement algorithm featuring explicit design space exploration. *ACM Trans. on Design Automation of Electronic Systems*, 2:62–80, 1997.

[58] H. Esbensen and E.S. Kuh. EXPLORER: an interactive floorplaner for design space exploration. In *European Design Automation Conf.*, pages 356–361, 1996.

[59] H. Esbensen and P. Mazumder. A genetic algorithm for the steiner problem in graphs. In *European Design & Test Conf.*, pages 402–406, 1994.

[60] H. Esbensen and P. Mazumder. SAGA: A unification of the genetic algorithm with simulated annealing and its application to macro-cell placement. In *VLSI Design Conf.*, pages 211–214, 1994.

[61] M. Escobar and F. Somenzi. Synthesis of AND-EXOR expressions via satisfyability. *IFIP WG 10.5 Workshop on Applications of the Reed-Muller Expansion in Circuit Design*, pages 80–87, 1995.

[62] S.C. Fang, W.S. Feng, and S.L. Lee. A new efficient approach to multilayer channel routing problem. In *Design Automation Conf.*, pages 579–584, 1992.

[63] P.J. Fleming and A.P. Pashkevich. Computer aided control system design using a multiobjective optimization approach. In *Control Conference*, pages 174–179, 1985.

[64] C.M. Fonseca and P.J. Fleming. Genetic algorithms for multiobjective optimization: Formulation, discussion and generalization. In *Int'l Conference on Genetic Algorithms*, pages 416–423, 1993.

[65] C.M. Fonseca and P.J. Fleming. An overview of evolutionary algorithms in multiobjective optimization. *Evolutionary Computation*, 3(1):1–16, 1995.

[66] S.J. Friedman and K.J. Supowit. Finding the optimal variable ordering for binary decision diagrams. In *Design Automation Conf.*, pages 348–356, 1987.

[67] H. Fujii, G. Ootomo, and C. Hori. Interleaving based variable ordering methods for ordered binary decision diagrams. In *Int'l Conf. on CAD*, pages 38–41, 1993.

[68] M. Fujita, H. Fujisawa, and N. Kawato. Evaluation and improvements of Boolean comparison method based on binary decision diagrams. In *Int'l Conf. on CAD*, pages 2–5, 1988.

[69] M. Fujita, H. Fujisawa, and Y. Matsunaga. Variable ordering algorithms for binary decision diagrams and their evolution. *IEEE Trans. on CAD*, 12:6–12, 1993.

[70] M. Fujita, Y. Matsunaga, and T. Kakuda. On variable ordering of binary decision diagrams for the application of multi-level synthesis. In *European Conf. on Design Automation*, pages 50–54, 1991.

[71] T. Gao, P.M. Vaidya, and C.L. Liu. A performance driven macro-cell placement algorithm. In *Design Automation Conf.*, pages 147–152, 1992.

[72] M.R. Garey and D.S. Johnson. *Computers and Intractability - A Guide to NP-Completeness.* Freemann, San Francisco, 1979.

[73] N. Göckel and R. Drechsler. Influencing parameters of evolutionary algorithms for sequencing problems. In *Int'l Conference on Evolutionary Computation*, pages 575–580, 1997.

[74] N. Göckel, R. Drechsler, and B. Becker. GAME: A software environment for using genetic algorithms in circuit design. In *Applications of Computer Systems*, 1997.

[75] N. Göckel, R. Drechsler, and B. Becker. Learning heuristics for OKFDD minimization by evolutionary algorithms. In *ASP Design Automation Conf.*, pages 469–472, 1997.

[76] N. Göckel, R. Drechsler, and B. Becker. A multi-layer detailed routing approach based on evolutionary algorithms. In *Int'l Conference on Evolutionary Computation*, pages 557–562, 1997.

[77] N. Göckel, M. Keim, R. Drechsler, and B. Becker. A genetic algorithm for sequential circuit test generation based on symbolic fault simulation. In *European Test Workshop*, 1997.

[78] N. Göckel, G. Pudelko, R. Drechsler, and B. Becker. A hybrid genetic algorithm for the channel routing problem. In *Int'l Symp. Circ. and Systems*, pages IV:675–IV:678, 1996.

[79] D.E. Goldberg. *Genetic Algorithms in Search, Optimization & Machine Learning.* Addision-Wesley Publisher Company, Inc., 1989.

[80] D.E. Goldberg and R. Lingle. Alleles, loci, and the traveling salesman problem. In *Int'l Conference on Genetic Algorithms*, pages 154–159, 1985.

[81] J. Hartmann. On numerical weight optimization for random test. In *European Conf. on Design Automation*, pages 223–230, 1993.

[82] J. Hartmann and G. Kemnitz. How to do weighted random testing for BIST? In *Int'l Conf. on CAD*, pages 568–571, 1993.

[83] J.H. Holland. *Adaption in Natural and Artifical Systems.* The University of Michigan Press, Ann Arbor, MI, 1975.

[84] M.S. Hsiao, E.M. Rudnick, and J.H. Patel. Alternating strategies for sequential circuit atpg. In *European Design & Test Conf.*, pages 368–374, 1996.

[85] M.D. Huang, F. Romeo, and A.L. Sangiovanni-Vincentelli. An efficient general cooling schedule for simulated annealing. In *Int'l Conf. on CAD*, pages 381–384, 1986.

[86] N. Ishiura, H. Sawada, and S. Yajima. Minimization of binary decision diagrams based on exchange of variables. In *Int'l Conf. on CAD*, pages 472–475, 1991.

[87] R. Joobbani. *An Artificial Intelligence Approach to VLSI Routing.* MA: Kluwer Academic Publishers, 1986.

[88] R. Kapur, S. Patil, T.J. Snethen, and T.W. Williams. Design of an efficient weighted random pattern generation system. In *Int'l Test Conf.*, pages 491–500, 1994.

[89] R.M. Karp. Reducibility among combinatorial problems. In R.E. Miller and J.W. Thatcher, editors, *Complexity of Computer Computations*, pages 85–103. 1972.

[90] U. Kebschull and W. Rosenstiel. Efficient graph-based computation and manipulation of functional decision diagrams. In *European Conf. on Design Automation*, pages 278–282, 1993.

[91] B.W. Kernighan. An optimum channel routing algorithm for polycell layouts of integrated circuits. In *Proceedings of the 10th Design Automation Workshop*, pages 50–59, 1973.

[92] V. Kommu and I. Pomeranz. GAFPGA: Genetic algorithm for FPGA technology mapping. In *European Design Automation Conf.*, pages 300–305, 1993.

[93] J. Koza. *Genetic Programming - On the Programming of Computers by means of Natural Selection.* MIT Press, 1992.

[94] T. Kozlowski, E. L. Dagless, and J. M. Saul. An enhanced algorithm for the minimization of exclusive-or sum-of-products for incompletely specified functions. In *Int'l Conf. on Comp. Design*, pages 244–249, 1995.

[95] R. Krieger. PLATO: A tool for computation of exact signal probabilities. In *VLSI Design Conf.*, pages 65–68, 1993.

[96] R. Krieger, B. Becker, and M. Keim. A hybrid fault simulator for synchronous sequential circuits. submitted to Journal of Electronic Testing, 1994.

[97] R. Krieger, B. Becker, and M. Keim. A hybrid fault simulator for synchronous sequential circuits. In *Int'l Test Conf.*, pages 614–623, 1994.

[98] R. Krieger, B. Becker, and C. Ökmen. OBDD-based optimization of input probabilities for weighted random test. In *Int'l Symp. on Fault-Tolerant Comp.*, pages 120–129, 1995.

[99] R. Krieger, B. Becker, and R. Sinković. A BDD-based algorithm for computation of exact fault detection probabilities. In *Int'l Symp. on Fault-Tolerant Comp.*, pages 186–195, 1993.

[100] B. Krishnamurthy and I.G. Tollis. Improved techniques for estimating signal probabilities. *IEEE Trans. on Comp.*, 38(7):1041–1045, 1989.

[101] A. Kuehlmann and L.P.P.P. van Ginneken. Grammar-based optimization of synthesis scenarios. In *Int'l Conf. on Comp. Design*, pages 20–25, 1994.

[102] E.L. Lawler. *Combinatorial Optimization: Networks and Matroids.* Holt, Rinehart and Winston, New York, 1976.

[103] T. Lengauer. *Combinatorial Algorithms for Integrated Circuit Layout.* Teubner, Wiley, 1990.

[104] J. Lienig. Channel and switchbox routing with minimized crosstalk - a parallel genetic algorithm approach. In *VLSI Design Conf.*, pages 27–31, 1997.

[105] J. Lienig. A parallel genetic algorithm for performance-driven VLSI routing. *IEEE Trans. on Evolutionary Comp.*, 1(1):29–39, 1997.

[106] J. Lienig and K. Thulasiraman. A genetic algorithm for channel routing in VLSI circuits. *Evolutionary Computation*, 1(4):293–311, 1993.

[107] J. Lienig and K. Thulasiraman. A new genetic algorithm for the channel routing problem. In *VLSI Design Conf.*, pages 133–136, 1994.

[108] J. Lillis, C.K. Cheng, and T.Y. Lin. Optimal wire sizing and buffer insertion for low power and a generalized delay model. In *Int'l Conf. on CAD*, pages 138–143, 1995.

[109] Y.-L. Lin, Y.-C. Hsu, and F.-S. Tsai. Silk: A simulated evolution router. *IEEE Trans. on CAD*, 8(10):1108–1114, 1989.

[110] S. Malik, A.R. Wang, R.K. Brayton, and A.L. Sangiovanni-Vincentelli. Logic verification using binary decision diagrams in a logic synthesis environment. In *Int'l Conf. on CAD*, pages 6–9, 1988.

[111] Z. Michalewicz. *Genetic Algorithms + Data Structures = Evolution Programs.* Springer-Verlag, 1994.

[112] Z. Michalewicz. Heuristic methods for evolutionary computation techniques. *Journal of Heuristics*, 1:177–206, 1995.

[113] A. Miczo. The sequential ATPG: A theoretical limit. In *Int'l Test Conf.*, pages 143–147, 1983.

[114] S. Minato, N. Ishiura, and S. Yajima. Shared binary decision diagrams with attributed edges for efficient Boolean function manipulation. In *Design Automation Conf.*, pages 52–57, 1990.

[115] D. Möller, P. Molitor, and R. Drechsler. Symmetry based variable ordering for ROBDDs. *IFIP Workshop on Logic and Architecture Synthesis, Grenoble*, pages 47–53, 1994.

[116] B.M.E. Moret and H.D. Shapiro. *Algorithms from P to NP*. The Benjamin/Cummings Publisher Company, 1991.

[117] R. Murgai, R.K. Brayton, and A.L. Sangiovanni-Vincentelli. *Logic Synthesis for Field-Programmable Gate Arrays*. Kluwer Academic Publisher, 1995.

[118] T.M. Niermann and J.H. Patel. HITEC: A test generation package for sequential circuits. In *European Conf. on Design Automation*, pages 214–218, 1991.

[119] Y. Nishizaki, M. Igusa, and A. Sangiovanni-Vincentelli. Mercury: A new approach to macro-cell global routing. In *Proceedings of the IFIP 10/WG 10.5 International Conference on VLSI*, 1989.

[120] K. Ohomori and T. Kasai. Logic synthesis using a genetic algorithm. In *Int'l Symposium on IC Technologies, Systems and Applications*, pages 200–203, 1997.

[121] C. Ökmen, M. Keim, R. Krieger, and B. Becker. On optimizing BIST architecture by using OBDD-based approaches and genetic algorithms. In *VLSI Test Symp.*, pages 426–431, 1997.

[122] S. Panda and F. Somenzi. Who are the variables in your neighborhood. In *Int'l Conf. on CAD*, pages 74–77, 1995.

[123] S. Panda, F. Somenzi, and B.F. Plessier. Symmetry detection and dynamic variable ordering of decision diagrams. In *Int'l Conf. on CAD*, pages 628–631, 1994.

[124] M.A. Perkowski and M. Chrzanowska-Jeske. An exact algorithm to minimize mixed-radix exclusive sums of products for incompletely specified Boolean functions. In *Int'l Symp. Circ. and Systems*, pages 1652–1655, 1990.

[125] I. Pomeranz and S. Reddy. On improving genetic optimization based test generation. In *European Design & Test Conf. User Forum*, 1997.

[126] I. Pomeranz and S.M. Reddy. 3-weight pseudo-random test generation based on a deterministic test set. *IEEE Trans. on Comp.*, pages 1040–1049, 1993.

[127] A.T. Rahmani and N. Ono. A genetic algorithm for the channel routing problem. In *Int'l Conference on Genetic Algorithms*, pages 494–498, 1993.

[128] I. Rechenberg. *Evolutionsstrategie.* Frommann-Holzboog, Stuttgart, FRG, 1973.

[129] S.M. Reddy. Easily testable realizations for logic functions. *IEEE Trans. on Comp.*, 21:1183–1188, 1972.

[130] B. Reeb and H.-J. Wunderlich. Deterministic pattern generation for weighted random pattern testing. In *European Design & Test Conf.*, pages 30–36, 1996.

[131] I.S. Reed. A class of multiple-error-correcting codes and their decoding scheme. *IRE Trans. on Inf. Theory*, 3:6–12, 1954.

[132] R. Rudell. Dynamic variable ordering for ordered binary decision diagrams. In *Int'l Conf. on CAD*, pages 42–47, 1993.

[133] E. Rudnick and J.H. Patel. A genetic approach to test application time reduction for full scan and partial scan circuits. In *VLSI Design Conf.*, 1995.

[134] E.M. Rudnick, J.H. Patel, G.S. Greenstein, and T.M. Niermann. Sequential circuit test pattern generation in a genetic algorithm framework. In *Design Automation Conf.*, pages 698–704, 1994.

[135] D.G. Saab, Y.G. Saab, and J.A. Abraham. Iterative [simulation-based genetics+deterministic techniques] =complete atpg. In *Int'l Conf. on CAD*, pages 40–43, 1994.

[136] A. Sarabi and M.A. Perkowski. Fast exact and quasi-minimal minimization of highly testable fixed-polarity AND/XOR canonical networks. In *Design Automation Conf.*, pages 30–35, 1992.

[137] A. Sarabi and M.A. Perkowski. Design for testability properties of AND/XOR networks. *IFIP WG 10.5 Workshop on Applications of the Reed-Muller Expansion in Circuit Design*, pages 147–153, 1993.

[138] T. Sasao. AND-EXOR expressions and their optimization. In T. Sasao, editor, *Logic Synthesis and Optimization*, pages 287–312. Kluwer Academic Publisher, 1993.

[139] T. Sasao. *Logic Synthesis and Optimization.* Kluwer Academic Publisher, 1993.

[140] T. Sasao and Ph. Besslich. On the complexity of mod-2 sum PLAs. *IEEE Trans. on Comp.*, 39:262–266, 1990.

[141] T. Sasao and F. Izuhara. Exact minimization of AND-EXOR expressions using multi-terminal EXOR ternary decision diagrams. In *IFIP WG 10.5 Workshop on Applications of the Reed-Muller Expansion in Circuit Design*, pages 213–220, 1995.

[142] J. Saul. Logic synthesis for arithmetic circuits using the Reed-Muller representation. In *European Conf. on Design Automation*, pages 109–113, 1992.

[143] J. Savir, G.S. Ditlow, and P.H. Bardell. Random pattern testability. *IEEE Trans. on Comp.*, 33:79–95, 1984.

[144] C. Scholl and P. Molitor. Communication based FPGA synthesis for multi-output Boolean functions. In *ASP Design Automation Conf.*, pages 279–287, 1995.

[145] C. Sechen. *VLSI Placement and Global Routing Using Simulated Annealing*. Kluwer Academic Publishers, Dordrecht, The Netherlands, 1988.

[146] H. Shin and A. Sangiovanni-Vincentelli. A detailed router based on incremental routing modifications: Mighty. *IEEE Trans. on CAD*, 6(6):942–955, 1987.

[147] T. Stanion and D. Bhattacharya. TSUNAMI: a path oriented scheme for algebraic test generation. In *Int'l Symp. on Fault-Tolerant Comp.*, pages 36–43, 1991.

[148] L. Stockmeyer. Optimal orientations of cells in slicing floorplan designs. *Information and Control*, 57:91–101, 1983.

[149] G.F. Sullivan. Approximation algorithms for steiner tree problems. Technical report, Technical Report 249, Dept. of Computer Science, Yale University, 1982.

[150] C.C. Tsai and M. Marek-Sadowska. Efficient minimization algorithms for fixed polarity AND/XOR canonical networks. In *Great Lakes Symp. VLSI*, pages 76–79, 1993.

[151] C.C. Tsai and M. Marek-Sadowska. Boolean matching using generalized Reed-Muller forms. In *Design Automation Conf.*, pages 339–344, 1994.

[152] C.C. Tsai and M. Marek-Sadowska. Logic synthesis for testability. In *Great Lakes Symp. VLSI*, pages 118–121, 1996.

[153] T.Sasao. An exact minimization of AND-EXOR expressions using BDDs. *IFIP WG 10.5 Workshop on Applications of the Reed-Muller Expansion in Circuit Design*, pages 91–98, 1993.

[154] R. Venkateswaran and P. Mazumder. Routing algorithms in semiconductor circuit design. In preparation, 1994.

[155] J.A. Waicukauski, E. Linbloom, E.B. Eichelberger, and O.P. Forlenza. A method for generating weighted random test patterns. *IBM J. Res. Develop.*, 33(2):149–161, 1989.

[156] D.F. Wong and C.L. Liu. A new algorithm for floorplan design. In *Design Automation Conf.*, pages 101–107, 1986.

[157] H.-J. Wunderlich. PROTEST: A tool for probabilistic testability analysis. In *Design Automation Conf.*, 1985.

[158] H.-J. Wunderlich. On computing optimized input probabilities for random tests. In *Design Automation Conf.*, pages 392–398, 1987.

[159] B. Wurth, K. Eckl, and K. Antreich. Functional multiple-output decomposition: Theory and implicit algorithm. In *Design Automation Conf.*, pages 54–59, 1995.

[160] S. Yang. Logic synthesis and optimization benchmarks user guide. Technical Report 1/95, Microelectronic Center of North Carolina, Jan. 1991.

Index